SHAW'S SENSE OF HISTORY

Shaw's Sense of History

J. L. WISENTHAL

CLARENDON PRESS · OXFORD
1988

Oxford University Press, Walton Street, Oxford OX2 6DP
Oxford New York Toronto
Delhi Bombay Calcutta Madras Karachi
Petaling Jaya Singapore Hong Kong Tokyo
Nairobi Dar es Salaam Cape Town
Melbourne Auckland
and associated companies in
Beirut Berlin Ibadan Nicosia

Oxford is a trade mark of Oxford University Press

Published in the United States
by Oxford University Press, New York

© Jonathan Wisenthal 1988
© Shaw Letters 1988
The Trustees of the British Museum,
the Governors and Guardians of the
National Gallery of Ireland, and
the Royal Academy of Dramatic Art

All rights reserved. No part of this publication may be reproduced,
stored in a retrieval system, or transmitted, in any form or by any means,
electronic, mechanical, photocopying, recording, or otherwise, without
the prior permission of Oxford University Press

British Library Cataloguing in Publication Data
Wisenthal, J. L.
Shaw's sense of history.
1. Shaw, Bernard—Criticism and
interpretation
I. Title
822'.912 PR5367
ISBN 0-19-812892-4

Library of Congress Cataloging in Publication Data
Data available

Set by Rowland Phototypesetting Ltd.
Printed in Great Britain by
Biddles Ltd.
Guildford and King's Lynn

For Stephen and Rosalind

Preface

'I AM fond of women (or one in a thousand, say)', Shaw explained to Ellen Terry in 1897; 'but I am in earnest about quite other things. To most women one man and one lifetime make a world. I require whole populations and historical epochs to engage my interests seriously.' The subject of my study is Shaw's engagement as a dramatist with whole populations and historical epochs; that is, his sense of history—past and present—as it expresses itself in the plays.

Ten of Shaw's plays have an obvious historical basis, in that they are set in the historical past. I want to show how an awareness of Shaw's historical attitudes can illuminate one's response to such works as *Caesar and Cleopatra*, *Saint Joan*, and *'In Good King Charles's Golden Days'*, and furthermore I want to show that many of Shaw's fifty-two plays can profitably be seen in the light of his thinking about history. Thus plays like *Man and Superman*, *John Bull's Other Island*, *Major Barbara*, *Heartbreak House*, and *Back to Methuselah*, though not conventionally history plays, occupy a significant place in this study. Apart from one or two special cases, it is not my concern to examine Shaw's treatment of sources; this has already been done for a number of the history plays by other scholars, and my aim is rather to demonstrate the importance of Shaw's interest in historical process, the design of history.

Such an interest sets Shaw in a Victorian context, and I see him as an essentially Victorian writer. He was 44 years old when Queen Victoria died, and his attitudes towards history are part of the Victorian intellectual world. I would like to think of my study as a continuation of work done by Julian B. Kaye, whose *Bernard Shaw and the Nineteenth-Century Tradition* was published in 1958, and particularly Martin Meisel, whose *Shaw and the Nineteenth-Century Theater* (1963) remains one of the outstanding pieces of critical writing about Shaw. Meisel places Shaw's plays in a Victorian dramatic tradition; I want to place them in a Victorian intellectual context, specifically that of Victorian historiography. That is why I have deliberately begun my introductory chapter, which provides the necessary background information for the rest of the enquiry, with a brief discussion of Carlyle, Macaulay, and some other Victorian figures. The chapters that follow will explore ways in which Shaw's plays move between the poles of Carlyle and Macaulay, between the antithetical Victorian intellectual traditions that these two historical writers may be seen to represent.

This is not a study of 'Shaw the Historian', for the good reason that Shaw was not a historian. He was a dramatist, and I have tried to keep the focus on his dramatic achievement. Shaw wrote so much splendid non-dramatic prose that it is a great temptation, to which I have generally yielded, to quote at length for purposes of illustration. Even prose that does not represent Shaw at his best, as in the late *Everybody's Political What's What?*, has often proved irresistible because of its illustrative value. But all the passages quoted from *The Intelligent Woman's Guide to Socialism*, *Everybody's Political What's What?*, and other non-dramatic prose writings, are there in order to make explicit the historical attitudes that underlie the plays. The object of my study is not to enhance Shaw's reputation as a thinker but to contribute towards a fuller understanding of what he is doing as a playwright.

Acknowledgements

QUOTATIONS from Shaw's writing are by permission of the Society of Authors on behalf of the Bernard Shaw Estate. The excerpt from Yeats's 'Two Songs from a Play' in Chap. 5 is by permission of A. P. Watt Ltd. on behalf of Michael B. Yeats and Macmillan London Ltd. I also wish to express my gratitude to Dan H. Laurence and Michael K. Goldberg.

Contents

Note on References	xii
1. Introduction	1
2. The Heroic in History	56
3. The Middle Ages, the Renaissance, and After	77
4. Progress an Illusion?	101
5. Present History	135
6. Shavian History and Shavian Drama	165
Index	181

Note on References

QUOTATIONS from Shaw's plays, prefaces, and certain related writings are taken from *The Bodley Head Bernard Shaw: Collected Plays with their Prefaces*, ed. Dan H. Laurence (7 vols., London: Max Reinhardt, The Bodley Head, 1970–4). Such quotations are indicated parenthetically in my text, with the abbreviation *CPP* and the volume number.

1
Introduction

i. A Preliminary Glance at Carlyle, Macaulay, and Others

On 4 April 1850, Macaulay recorded his opinion of Carlyle's newly published *Latter-Day Pamphlets*. He was not greatly impressed. 'I read Carlyle's Trash—Latterday something or other—beneath criticism', he recorded in his journal. 'Surely the world will not be duped for ever by such an empty headed bombastic dunce.'[1] Eight years later he registered a similar response to *Frederick the Great*: 'I never saw a worse book. Nothing new of the smallest value. The philosophy nonsense and the style gibberish.'[2] Carlyle's judgements on Macaulay were hardly more flattering. In the early 1830s, when both men were beginning to make their name, Carlyle observed the 'force and emphasis' in his contemporary, but he noted in his journal: 'Wants the root of belief, however. May fail to accomplish much.' Another journal entry pointed to the same failing: 'The strongest young man, one Macaulay (now in Parliament, as I from the first predicted), an emphatic, hottish, really forcible person, but unhappily without divine idea.'[3] When Macaulay's *History of England* was being widely celebrated in England, Carlyle's opinion of the work was decidedly unfavourable. 'Four hundred editions could not lend it any permanent value', he wrote, 'there being no depth of sense in it, and a very great quantity of rhetorical wind.'[4] When in 1867 he deplored the fact that

---

[1] Quoted in Joseph Hamburger, *Macaulay and the Whig Tradition* (Chicago: Univ. of Chicago Press, 1976), 258 n. 35.

[2] Quoted in Hamburger, 176 (Macaulay's Journal, 30 Sept. 1858). Macaulay was responding to the early volumes of *Frederick*; he died before the later ones were published.

[3] Quoted in James Anthony Froude, *Thomas Carlyle: A History of the First Forty Years of His Life 1795–1835* (2 vols., London: Longmans, Green, 1882), ii. 373, 231. Cf. Carlyle's letter to his mother, 14 Sept. 1831: 'Of Macaulay I hear nothing very good—a sophistical, rhetorical, ambitious young man of talent; "set in there," as Mill said, "to make flash speeches, and he makes them." It seems to me of small consequence whether we meet at all' (ibid. 201).

[4] Quoted in G. P. Gooch, *History and Historians in the Nineteenth Century* (1913; London: Longmans, 1952), 285.

England's history 'that is Divine' had 'fallen into such a set of hands!' there can be little doubt that Macaulay's were among the hands he had in mind. The interpretation of England's history 'in the present ages', he complained, is 'scandalously ape-like, I must say; impious, blasphemous;—totally incredible withal.'[5]

In spite of this mutual antagonism, there was a good deal that Carlyle and Macaulay had in common. They were the two leading 'literary' historians in nineteenth-century England, and their writing careers were contemporaneous—roughly the 1820s to the 1850s. Both were Victorian men of letters, who first achieved prominence in the quarterlies; and the two of them were attracted to many of the same subjects. Carlyle's *Oliver Cromwell's Letters and Speeches*—one of his major pieces of historical work—and Macaulay's *History of England* are both studies of English history of the seventeenth century. Carlyle's lengthy biography of Frederick the Great has a minor counterpart in Macaulay's essay on Frederick. Similarly, one can set Macaulay's essays on Mirabeau and Barère next to Carlyle's *The French Revolution*.

As historians, Carlyle and Macaulay wrote about the past with the Victorian present very much in mind. Macaulay would have agreed with Carlyle's comment about seventeenth-century England: '[T]hus do the two centuries stand related to me—the seventeenth *worthless* except precisely in so far as it can be made the *nineteenth*.'[6] Each of them wrote about the past in order to provide inspiration and direction for their contemporaries. And like many historians in Victorian England (Froude, for example), Carlyle and Macaulay are notably pro-Protestant in their outlook. Even though Carlyle uses certain aspects of the Catholic Middle Ages to reveal the degradation of Victorian England, his main sympathies are with heroes of Protestantism; his Frederick and Macaulay's William III have at least this one characteristic in common: they are presented as defenders of Protestant Europe against Catholic France and its allies.

In spite of these similar interests, however, there is really no

---

[5] Thomas Carlyle, 'Shooting Niagara: and After?', *Critical and Miscellaneous Essays*, v. 27, *The Works of Thomas Carlyle*, ed. H. D. Traill (Centenary Edn., 30 vols., London: Chapman and Hall, 1896–9), xxx.

[6] Carlyle to R. W. Emerson, 29 Aug. 1842, quoted in René Wellek, 'Carlyle and the Philosophy of History', *Philological Quarterly*, 23 (Oct. 1944), 70.

cause for wonder (as Macaulay would say) that these two men took such a dim view of each other. One could explore at great length their contrasting treatments of the same subjects—their respective judgements of Frederick, for example—but for our purposes we might look briefly at just three areas: their views of the Great Man, of progress, and of the direction of English history.

Both Carlyle and Macaulay have their heroes. The main hero of Macaulay's *History of England* is William III, and there are secondary heroes such as Lord Somers. The qualities that Macaulay most admires in Lord Halifax (1633–95) provide a good expression of some of the central values in the *History*. Halifax 'was the chief of those politicians whom the two great parties contemptuously called Trimmers', Macaulay tells us in introducing Charles II's ministry of 1679. 'Instead of quarrelling with this nickname, he assumed it as a title of honour, and vindicated, with great vivacity, the dignity of the appellation.' As a Trimmer on principle and by temperament, he avoided the political extremes. 'His place was on the debatable ground between the hostile divisions of the community, and he never wandered far beyond the frontier of either.'[7] The Marquis of Halifax would not have been one of Carlyle's heroes, and to compare Macaulay's Halifax, William, or Somers with Carlyle's Frederick, Cromwell, or Mahomet is to see something of the distance between these two Victorian men of letters. For Macaulay the essence of politics is compromise. The essential characteristic of Carlyle's heroes is their refusal to compromise. They are the leading combatants on the truthful, divine side in 'that great universal war which alone makes-up the true History of the World,—the war of Belief against Unbelief'.[8]

Carlyle's heroes are in touch with the profound truths of the universe, and any failings are therefore insignificant. Carlyle makes absolute distinctions between the hero and the scoundrel, the genuine man and the sham, the ablest man in England and the incompetents. He applies to public affairs the Calvinist distinction between the saved and the damned. In his book on Cromwell he may at times pity Cromwell's opponents, but he consistently

---

[7] T. B. Macaulay, *The History of England from the Accession of James II*, *The Works of Lord Macaulay*, ed. Lady Trevelyan (8 vols., London: Longmans, Green, 1879), i. 192 (chap. II).
[8] Carlyle, *On Heroes, Hero-Worship, and the Heroic in History*, *Works*, v. 204.

endorses everything that Cromwell says or does. Cromwell sets his standard for him, and he accepts all of Cromwell's assumptions. Macaulay, on the other hand, thrives on human inconsistency. His portrait of William may be more favourable than most historians would accept, but he does acknowledge failings on the part of his hero, and there is no core of intense heroic soul that renders these human limitations insignificant. In his essay on Warren Hastings, he says that 'It is good to be often reminded of the inconsistency of human nature, and to learn to look without wonder or disgust on the weaknesses which are found in the strongest minds'[9]—and one of the examples he offers, as it happens, is that of Frederick the Great, the hero on whom Carlyle lavishes his fullest treatment.

The most striking difference between Carlyle and Macaulay as historians lies in their contrasting attitudes towards progress. 'Between the middle of the last century and 1914', writes E. H. Carr, 'it was scarcely possible for a British historian to conceive of historical change except as change for the better.'[10] The historian who most clearly represents this commitment to the idea of progress is Macaulay. His commitment, and indeed his enthusiasm, are unqualified. One does not need to proceed far in his writings before coming across a passage that could stand as an extreme example of the Victorian belief in progress. One of the better known is in his *Edinburgh Review* essay on Lord Bacon.

Ask a follower of Bacon what the new philosophy, as it was called in the time of Charles the Second, has effected for mankind, and his answer is ready; 'It has lengthened life; it has mitigated pain; it has extinguished diseases; it has increased the fertility of the soil; it has given new securities to the mariner; it has furnished new arms to the warrior; it has spanned great rivers and estuaries with bridges of form unknown to our fathers; it has guided the thunderbolt innocuously from heaven to earth; it has lighted up the night with the splendour of the day; it has extended the range of the human vision; it has multiplied the power of the human muscles; it has accelerated motion; it has annihilated distance; it has facilitated intercourse, correspondence, all friendly offices, all despatch of business; it has enabled man to descend to the depths of the sea, to soar into the air, to penetrate securely into the noxious recesses of the earth, to traverse the land in cars which whirl along without horses, and

---

[9] Macaulay, 'Warren Hastings', *Works*, vi. 641–2.
[10] E. H. Carr, *What Is History?* (London: Macmillan, 1961), 32.

the ocean in ships which run ten knots an hour against the wind. These are but a part of its fruits, and of its first fruits. For it is a philosophy which never rests, which has never attained, which is never perfect. Its law is progress. A point which yesterday was invisible is its goal to-day, and will be its starting-post to-morrow.'[11]

This passage does not give the best impression of Macaulay's writing; it is the kind of passage one might produce in an attempt to discredit him as 'the great apostle of the Philistines' (in Matthew Arnold's phrase). But it does represent an attitude that is fundamental to his writing, and that was widespread—though not universal—in Victorian England. It is a passage that supports R. G. Collingwood's acerbic generalization that 'In the later nineteenth century the idea of progress became almost an article of faith.... The progress of humanity, from the nineteenth-century point of view, meant getting richer and richer and having a better and better time.'[12] Progress in this passage of Macaulay's means *material* progress.

Another passage that could be used to discredit Macaulay (not my intention here, needless to say) occurs in another of his *Edinburgh Review* essays from the 1830s. This time he is writing not about the advancement of mankind in general, but—characteristically—about the advancement of his own nation.

The history of England is emphatically the history of progress. It is the history of a constant movement of the public mind, of a constant change in the institutions of a great society. We see that society, at the beginning of the twelfth century, in a state more miserable than the state in which the most degraded nations of the East now are. We see it subjected to the tyranny of a handful of armed foreigners. We see a strong distinction of caste separating the victorious Norman from the vanquished Saxon. We see the great body of the population in a state of personal slavery. We see the most debasing and cruel superstition exercising boundless dominion over the most elevated and benevolent minds. We see the multitude sunk in brutal ignorance, and the studious few engaged in acquiring what did not deserve the name of knowledge. In the course of seven centuries the wretched and degraded race have become the greatest and most highly civilised people that ever the world saw, have spread their dominion over every quarter of the globe, have scattered the seeds of mighty empires

---

[11] Macaulay, 'Lord Bacon', *Works*, vi. 222–3.
[12] R. G. Collingwood, *The Idea of History* (1946; London: Oxford Univ. Press, 1977), 144.

and republics over vast continents of which no dim intimation had ever reached Ptolemy or Strabo, have created a maritime power which would annihilate in a quarter of an hour the navies of Tyre, Athens, Carthage, Venice, and Genoa together, have carried the science of healing, the means of locomotion and correspondence, every mechanical art, every manufacture, every thing that promotes the convenience of life, to a perfection which our ancestors would have thought magical, have produced a literature which may boast of works not inferior to the noblest which Greece has bequeathed to us, have discovered the laws which regulate the motions of the heavenly bodies, have speculated with exquisite subtilty on the operations of the human mind, have been the acknowledged leaders of the human race in the career of political improvement. The history of England is the history of this great change in the moral, intellectual, and physical state of the inhabitants of our own island. There is much amusing and instructive episodical matter; but this is the main action.[13]

Here the progress is not only material but also moral and intellectual; and as in the passage on Bacon's inductive philosophy, the confident, relentless, cadences of the prose reflect the forward and upward development of a society.

In order to see that this ardent belief in improvement is not an isolated phenomenon in Victorian England, we might look for a moment at the work of one of Macaulay's contemporaries, Henry Thomas Buckle. Buckle has had the misfortune to be remembered —in so far as he is remembered at all—only as a peripheral Victorian figure. But there was a time in the nineteenth century when his name was illustrious. In 1857, when the first volume of his *History of Civilization in England* appeared, he was lionized as one of the foremost thinkers of the age. He died in Damascus five years later, and through the rest of the century his *History of Civilization* was read and discussed, not only in England but in the United States and (in translation into various languages) in Europe. When one character in Chekhov's *The Cherry Orchard* (1904) asks another, 'Have you read Buckle?', this would not have seemed an odd question at the time.

Buckle, like Macaulay, had great faith in the power of inductive scientific reasoning to improve the human lot; and like Macaulay, he saw the veneration for the past as a pernicious delusion. 'A people who regard the past with too wistful an eye, will never bestir

---

[13] Macaulay, 'Sir James Mackintosh', *Works*, vi, 95–6.

themselves to help the onward progress; they will hardly believe that progress is possible,' he wrote of the Spaniards.[14] The idea that progress *is* possible underlies the *History of Civilization*. One of Hume's limitations as a historian, according to Buckle, is that he lacked 'that invaluable quality of imagination without which no one can so transport himself into past ages as to realize the long and progressive movements of society, always fluctuating, yet, on the whole, steadily advancing.'[15] As in Macaulay's writings, the chief example of this historical tendency is post-Reformation England. Most twentieth-century readers would feel some sympathy with Leslie Stephen's suggestion in 1880 that Buckle is representative of 'that curious tone of popular complacency which was prevalent some thirty years ago, when people held that the devil had finally committed suicide upon seeing the Great Exhibition, having had things pretty much his own way till Luther threw the inkstand in his face'.[16]

One other indication of Macaulay's representative nature is an early poem of Tennyson's, 'You Ask Me Why', written in the 1830s. The question posed in the title is why 'Within this region I subsist'. The answer:

> It is the land that freemen till,
>   That sober-suited Freedom chose,
>   The land, where girt with friends or foes
> A man may speak the thing he will;
>
> A land of settled government,
>   A land of just and old renown,
>   Where Freedom slowly broadens down
> From precedent to precedent:
>
> Where faction seldom gathers head,
>   But by degrees to fulness wrought,
>   The strength of some diffusive thought
> Hath time and space to work and spread.[17]

These passages from Macaulay, Buckle, and Tennyson reflect

---

[14] H. T. Buckle, *History of Civilization in England* (1857–61; 3 vols., London: Longmans, Green, 1878), ii. 595 (chap. VIII).
[15] Buckle, *History of Civilization*, iii. 331 (chap. V).
[16] Leslie Stephen, 'An Attempted Philosophy of History', *Fortnightly Review*, 27 (1880), 673; quoted in Giles St Aubyn, *A Victorian Eminence: The Life and Works of Henry Thomas Buckle* (London: Barrie, 1958), 159.
[17] *The Poems of Tennyson*, ed. Christopher Ricks (London: Longmans, 1969), 490.

the mainstream of Victorian thinking on the subject of progress. We might bear in mind, too, the publication date of Darwin's *Origin of Species*: 1859, the year of Macaulay's death, and two years after the appearance of Buckle's first volume. Darwin's view of Nature as essentially progressive is one of the attitudes that marks his book as an expression of its age.

There were some Victorian voices, however, which expressed a very different sense of historical development. The most notable, and the most influential, of the opponents of the Victorian faith in progress was Carlyle.

For Carlyle there is constant change in history, but no linear pattern of forward, upward development. He looks closely at particular historical events and periods to reveal the workings of 'divine' revelation, justice, and retribution. In this pattern of recurrence, a nation parts company with fact, loses touch with the fundamental truths of the universe, and the inevitable result is a deserved destruction, sooner or later. History is the story of the various ways in which individuals and nations have responded to Nature and Fact. It is more a series of stories than a single coherent narrative.

Carlyle gives the impression that men have on the whole responded more effectively to Nature and Fact in the past than in the present, and he fiercely resists the Macaulayite perception that the natural direction of history is forward and upward. A past idea (such as Odin-worship) that was 'an honest insight into God's truth on man's part' does not die to be supplanted by newer and better ideas; it '*has* an essential truth in it which endures through all changes, an everlasting possession for us all'. The contrary, progressive view is deplorable:

[W]hat a melancholy notion is that, which has to represent all men, in all countries and times except our own, as having spent their life in blind condemnable error, mere lost Pagans, Scandinavians, Mahometans, only that we might have the true ultimate knowledge! All generations of men were lost and wrong, only that this present little section of a generation might be saved and right. They all marched foward there, all generations since the beginning of the world, like the Russian soldiers into the ditch of Schweidnitz Fort, only to fill-up the ditch with their dead bodies, that we might march-over and take the place! It is an incredible hypothesis.[18]

[18] Carlyle, *Heroes and Hero-Worship*, *Works*, v. 119–20 ('The Hero as Priest').

There is no explicit reference here to Macaulay, but the passage is one that irresistibly brings him to mind.

These opposing views of intellectual, moral, social, and material progress on the part of Macaulay and Carlyle lead to radically different assessments of the place of nineteenth-century England in relation to previous history. We could combine accuracy with simplicity by saying that for Macaulay the Victorian age is the best of times, while for Carlyle it is the worst of times. Macaulay's *History* frequently asserts the superiority of the present over the past. It acknowledges that progress will continue beyond the splendid heights attained by Victorian industrial civilization, but it exalts present achievements over anything that has gone before. The best-known section of the *History* is the third chapter, on the state of England in 1685. A recurring theme in this chapter is that English life in 1685 is inferior to English life in the 1840s. One might have thought that such an issue as child labour would subdue the force of this argument. Instead, Macaulay points to the prevalence of child labour in the seventeenth century, and he characteristically adds: 'The more carefully we examine the history of the past, the more reason shall we find to dissent from those who imagine that our age has been fruitful of new social evils. The truth is that the evils are, with scarcely an exception, old. That which is new is the intelligence which discerns and the humanity which remedies them.' He then devotes the last few pages of his chapter to an argument that 'the public mind of England has softened while it has ripened'; the English have in the course of ages become 'not only a wiser, but also a kinder people'.[19]

This outlook, in the 1840s and 1850s, was not of course unique to Macaulay. The idea that Victorian industrial civilization represents the culminating point of historical development thus far can be found in many other works of the period, including Buckle's *History of Civilization*. There is a footnote of Buckle's that expresses this attitude in an extreme and therefore striking way. He is talking about the Jacobite rebellion of 1745, which he sees as 'in our country, the last struggle of barbarism against civilization', and he quotes one of the Jacobites (i.e. barbarians) as recording that 'In every place we passed through, we found the

---

[19] Macaulay, *History of England*, *Works*, i. 327–31 (chap. III).

English very ill disposed towards us, except at Manchester, where there appeared some remains of attachment to the house of Stuart.' Buckle's comment on this is as follows: 'The champion of arbitrary power would find a different reception now, in that magnificent specimen of English prosperity, and of true, open-mouthed, English fearlessness. But a century ago, the men of Manchester were poor and ignorant.'[20]

This is the sort of sentiment that drew Carlyle's most strenuous denunciations. To him, the civilization represented by Manchester was the nadir, not the zenith, of historical development, and much of his finest writing is devoted to the exposure of the values of Manchester as false, inhuman, and cruel. Carlyle was contemptuous of 'the whole projecting, railwaying, knowledge-diffusing, march-of-intellect and otherwise promotive and locomotive societies in the Old and New World'.[21] *Past and Present* (1843) offers a telling contrast between twelfth-century monastic life and nineteenth-century industrial life. In the chapter entitled 'Twelfth Century', Carlyle celebrates the effectiveness of the feudal aristocracy that flourished then, and he intimates the transformation of England that lay in the future.

> The Ribble and the Aire roll down, as yet unpolluted by dyers' chemistry; tenanted by merry trouts and piscatory otters; the sunbeam and the vacant wind's-blast alone traversing those moors. Side by side sleep the coal-strata and the iron-strata for so many ages; no Steam-Demon has yet risen smoking into being. Saint Mungo rules in Glasgow; James Watt still slumbering in the deep of Time. *Mancunium*, Manceaster, what we now call Manchester, spins no cotton,—if it be not *wool* 'cottons,' clipped from the backs of mountain sheep. The Creek of the Mersey gurgles, twice in the four-and-twenty hours, with eddying brine, clangorous with sea-fowl; and is a *Lither*-Pool, a *lazy* or sullen Pool, no monstrous pitchy City, and Seahaven of the world![22]

Macaulay, regarding the present as the best of times, sees it as the culmination of progress from the barbarous and superstitious Middle Ages. Carlyle, seeing the present as the worst of times, suggests a decline from the Middle Ages.

---

[20] Buckle, *History of Civilization*, iii. 157, 159 n. (chap. III).
[21] Carlyle, 'Sir Walter Scott', *Critical and Miscellaneous Essays*, iv. 27, *Works*, xxix.
[22] Carlyle, *Past and Present*, *Works*, x. 66 (Book ii, chap. v).

Some of the contrasts between Macaulay and Carlyle may be seen in their respective treatments of seventeenth-century England. To Macaulay, the victory of English Protestantism in 1688, which is at the heart of the *History of England*, is the decisive moment in English history that lays the groundwork for nineteenth-century achievements. '[T]he history of our country during the last hundred and sixty years', he asserts in introducing his work, 'is eminently the history of physical, of moral, and of intellectual improvement.'[23] These words were published in 1848, and so 'the last hundred and sixty years' would take one back to 1688. Carlyle's attraction to the seventeenth century is of a very different kind. He does not consider William of Orange to be part of a progressive movement that leads from the Middle Ages to the present; he regards Cromwell as a divinely inspired hero who was part of an age of faith in the 1640s and 1650s—an age of faith that links the mid-seventeenth century with the Middle Ages. The purpose of Carlyle's *Oliver Cromwell's Letters and Speeches* is to make the unheroic nineteenth century aware of one of England's heroic ages: 'that Protectorate of Oliver Cromwell's', he asserted in a speech in 1866, 'appears to me to have been, on the whole, the most salutary thing in the modern history of England.'[24]

Another distinction between their responses to the seventeenth century is that whereas Macaulay's revolution is the orderly constitutional change of 1688, Carlyle is attracted to the violent disruption of the Civil War (as he was attracted to—at the same time that he was appalled by—the French Revolution). Macaulay gives close and sympathetic attention to party politics and parliamentary debates. Carlyle looks approvingly at Cromwell's decisive dismissal of Parliament and at his military victories. But the main difference between their attitudes towards seventeenth-century England is that whereas Macaulay sees the 1688 Revolution as the beginning of a continuing process that culminates in the Reform Bill of 1832, Carlyle sees Cromwell's regime as a period of revelation which stands in reproachful contrast to the unbelieving present. Carlyle greatly values Protestant heroes such as Cromwell and Luther, but he does not see the Reformation and

---

[23] Macaulay, *History of England*, Works, i. 2 (chap. 1).
[24] Carlyle, 'Inaugural Address at Edinburgh', *Critical and Miscellaneous Essays*, Works, iv. 459.

the seventeenth century as part of a teleological pattern, with the present as a point of culmination. What Carlyle most decisively rejects is the idea that 'the history of our country during the last hundred and sixty years is eminently the history of physical, of moral, and of intellectual improvement'. The physical improvement he would in part concede: 'The Practical Labour of England', he writes, is 'the one God's Voice we have heard in these two atheistic centuries.'[25] But the condition of the people is still wretched, and in any case the atheistic nature of the two centuries far outweighs the significance of any practical achievements. The main legacy of the seventeenth century is the godless era that was inaugurated by the Restoration of the monarchy in 1660. As opposed to Macaulay's positive connection between 1688 and 1832, we have Carlyle's pernicious connection between 1660 and the Second Reform Bill of 1867. Denouncing the Reform Bill in 1867, Carlyle commented: 'It has lasted long, that unblessed process; process of "lying to steep in the Devil's Pickle," for above two hundred years (I date the formal beginning of it from the year 1660, and desperate *return* of Sacred Majesty after such an ousting as it had got); process which appears to be now about complete.'[26] This is the period that almost coincides with Macaulay's 160 years of physical, of moral, and of intellectual improvement.

These three issues—the Great Man, progress, and the direction of English history—were large issues in the Victorian intellectual world. And they are large issues in the works of Bernard Shaw, who was a part of that world. In the Victorian divisions reflected in the writings of Carlyle and Macaulay, we can find some of the main concerns of Shaw's plays.

ii. *Shaw's Knowledge of Historians and History*

When we reflect on the fact that leading Victorian men of letters like Macaulay and Carlyle were mainly historians, we learn something about the state of Victorian culture. Nietzsche found 'nothing so remarkable in the man of the present day as his

[25] Carlyle, *Past and Present, Works,* x. 169 (Book iii, chap. vi).
[26] Carlyle, 'Shooting Niagara: and After?', *Critical and Miscellaneous Essays,* v. 14, *Works,* xxx.

Introduction   13

peculiar virtue and sickness called "the historical sense"',[27] and Jerome Hamilton Buckley's fine study, *The Triumph of Time*, has demonstrated just how much attention Victorian writers devoted to concepts of time, history, progress, and decadence.[28] One of the many Victorian writers cited by Buckley is William Morris, who talked about his 'passion for the history of the past of mankind', and who observed, '[T]he world has been noteworthy for more than one century and one place, a fact which we are pretty much apt to forget.'[29] Morris's writings often look to the past, especially the medieval past, and when in *News from Nowhere* he writes about a utopia in the future, his future is constructed as a re-creation of an imagined medieval past. Buckley notes that in the Victorian era 'almost every one of the major novelists attempted at least one historical novel, somehow relevant in theme to their own age, while many of the lesser ones peered into a more or less utopian future'.[30] At the same time, a number of the most popular serious plays on the Victorian stage were historical dramas; some of Henry Irving's more notable roles were in plays about Charles I, Richelieu, Louis XI, and Becket.

Tennyson said that his historical dramatic trilogy, consisting of *Becket*, *Harold*, and *Queen Mary*, 'pourtrays the making of England', and the *persona* in one of his late poems recollects that:

> oft
> On me, when boy, there came what then I called,
> Who knew no books and no philosophies,
> In my boy-phrase 'The Passion of the Past.'
>
> (ll. 216–19.)[31]

---

[27] Quoted in Peter Allan Dale, *The Victorian Critic and the Idea of History* (Cambridge, Mass.: Harvard Univ. Press, 1977), epigraph.
[28] Jerome Hamilton Buckley, *The Triumph of Time* (Cambridge, Mass.: Harvard Univ. Press, 1966). A recent discussion of the Victorian preoccupation with historical issues is A. Dwight Culler's *The Victorian Mirror of History* (New Haven and London: Yale Univ. Press, 1985).
[29] William Morris, 'How I Became a Socialist', 'Some Hints on Pattern-Designing', quoted in Buckley, 14, 16–17.
[30] Buckley, 3.
[31] [Hallam Tennyson], *Alfred Lord Tennyson: A Memoir* (2 vols., London: Macmillan, 1897), ii. 173; 'The Ancient Sage', *The Poems of Tennyson*, ed. Christopher Ricks, 1355.

Henry Kozicki has written a book on Tennyson's 'Passion of the Past', which records that in Tennyson's childhood home historical works such as David Hume's *History of England* and William Robertson's *History of Scotland* were used as family entertainment, and that Tennyson as an adult read the American historian John Lothrop Motley's *Rise of the Dutch Republic* aloud to his wife.[32] It was a sign of the times that chairs of history were established in Oxford and Cambridge in the 1860s, and when we think about the intellectual life of Victorian England it is well to remember that a historian like Macaulay was part of the educated person's equipment; an acquaintance with major historical writings would be taken for granted. One can see why Macaulay's great-nephew, G. M. Trevelyan, considered the Victorian age to have been 'the period when history in England reached the height of its popularity and of its influence on the national mind'.[33]

It was in this environment that Shaw lived. Having been born in Dublin in 1856, and having moved to London twenty years later, he shared his age's interest in time and in history. His historical reading began when he was a schoolboy. Although (according to his own account) he learned nothing from the curriculum, which included 'a pretence of English history ... (mostly false and scurrilous)',[34] he took care to educate himself outside school hours. Though nothing would have induced him 'to read the budget of stupid party lies that served as a text-book of history in school', he remembered reading Robertson's *Charles the Fifth* and his *History of Scotland* 'from end to end most laboriously' (Preface to *Misalliance*, *CPP* iv. 71). In his music criticism in the early 1890s, there are two references to a passage in Robertson's account of Mary Queen of Scots,[35] suggesting that he retained something of this early reading.

But Shaw claimed that his early historical education derived mainly from literary works. He noted in one of his music reviews that the Duke of Marlborough had acquired his knowledge of

[32] Henry Kozicki, *Tennyson and Clio: History in the Major Poems* (Baltimore: The Johns Hopkins Univ. Press, 1979), 3.
[33] George Macaulay Trevelyan, *Clio, A Muse and Other Essays* (London: Longmans, Green, 1930), 166.
[34] Shaw, *Sixteen Self Sketches* (London: Constable, 1949), 20–1.
[35] See *Shaw's Music*, ed. Dan H. Laurence (3 vols., London: Max Reinhardt, The Bodley Head, 1981), ii. 172, 980–1.

history from Shakespeare,[36] and Shaw was proud of the fact that as a schoolboy he had acquired most of his own early historical knowledge in a similar fashion. He claimed to have learnt French history from the novels of Dumas *père*, and English history from Shakespeare and Sir Walter Scott, and he said that he acquired his first acquaintance with the French Revolution when he was very young from Dickens's *A Tale of Two Cities*.[37]

In a late work, *Everybody's Political What's What?*, Shaw discussed the way in which history ought to be learned, drawing on his own experience:

Take the case of history as an indispensable part of the education of a citizen. Have you ever reflected on the impossibility of learning history from a collection of its bare facts in the order in which they actually occurred? You might as well try to gather a knowledge of London from the pages of the telephone directory. French history was not one of my school subjects; but by reading with great entertainment the historical novels of Dumas *père* I had a vivid conspectus of France from the sixteenth to the eighteenth century, from Chicot to Cagliostro, from the conquest of the nobility by the monarchy under Richelieu to the French Revolution. Like Marlborough, I had already learnt all I knew of English history, from King John to the final suicide of the English feudal aristocracy and its supersession by the capitalists on Bosworth field, from the chronicle plays of Shakespear. Adding to these congenial authorities the Waverley novels of Walter Scott I came out with a taste for history and an acquaintance with its personages and events which made the philosophy of history real for me when I was fully grown. Macaulay did not repel me as a historian (of sorts) nor Hegel and Marx bore and bewilder me.

In school he learnt very little history while his teacher was questioning the pupils about chapters in *The Students' Hume*, 'whereas when I was at home reading Quentin Durward, A Tale of Two Cities, or The Three Musketeers, I was learning it very agreeably'.[38]

As this passage suggests, Sir Walter Scott was an important source of Shaw's early knowledge of history. Not that all of Shaw's references to Scott are admiring; there is the notorious

---

[36] *Shaw's Music*, ii. 677–8.
[37] *Shaw's Music*, i. 43; ii. 892.
[38] Shaw, *Everybody's Political What's What?* (London: Constable, 1944), 180–1, 183.

review of *Cymbeline* in 1896, in which Scott is glancingly struck during an onslaught on Shakespeare: 'With the single exception of Homer, there is no eminent writer, not even Sir Walter Scott, whom I can despise so entirely ... when I measure my mind against his.'[39] Yet Shaw was evidently an avid reader of Scott's novels as a child, and in the year before his death in 1950 he introduced Rob Roy as a character in his puppet play, *Shakes versus Shav*. To Shakes's challenge, 'Couldst write Macbeth?', Shav replies,

> No need. He has been bettered
> By Walter Scott's Rob Roy. Behold, and blush.

Shakes then suffers the indignity of decapitation at the hands of Scott's representative (*CPP* vii. 474).[40] It was almost ninety years before writing this that Shaw had begun his historical education by reading Scott's novels, along with Shakespeare's history plays—and also Bunyan's *Pilgrim's Progress*, which offers a paradigm of historical development as meaningful divine pattern.

Another literary source of Shaw's historical knowledge was the many nineteenth-century plays and operas with historical settings. Bulwer-Lytton's *Richelieu*, for example, was familiar to him (the title role was one of the best-known roles of Barry Sullivan, the favourite actor of his boyhood),[41] and in 1922 he retained sharp recollections of Adelaide Ristori as Mary Stuart in Schiller's play: 'In my youth I saw the great Italian actress Ristori play Mary Stuart; and nothing in her performance remains more vividly with me than her representation of the relief of Mary at finding herself in the open air after months of imprisonment.'[42] Henry Irving was not among Shaw's most revered actors, but Shaw regarded Charles I, in W. G. Wills's play of that name, and

---

[39] Shaw, *Our Theatres in the Nineties* (3 vols., London: Constable, 1954), ii. 195.
[40] In a letter to St John Ervine, 5 Feb. 1941, Shaw praised Scott as 'the nearest thing to Shakespear that these islands have produced. He always gets there sooner or later with a stroke that settles the question.' He then quoted from memory a passage from Rob Roy that he later used in *Shakes versus Shav* (MS in Harry Ransom Humanities Research Center, the University of Texas at Austin). Ervine wrote in his biography of Shaw, 'Vehemently adverse to Walter Scott in his early years, he became devoted to him later in life' (*Bernard Shaw: His Life, Work and Friends* (London: Constable, 1956), 90).
[41] Shaw, Preface to *Ellen Terry and Bernard Shaw: A Correspondence*, ed. Christopher St John (London: Constable, 1931), p. xxvi.
[42] Shaw, 'Imprisonment', in *Doctors' Delusions, Crude Criminology, and Sham Education* (London: Constable, 1950), 177.

Louis XI in Dion Boucicault's play as two of Irving's most successful roles.[43] One history play that Shaw experienced as a reader rather than playgoer was Ibsen's *Emperor and Galilean* (subtitled 'A World-Historic Drama'), and his treatment of it in *The Quintessence of Ibsenism* makes it clear that this dramatization of Julian the Apostate was an important influence on him. And then there are nineteenth-century operas with historical settings, such as Bellini's *I Puritani* and Meyerbeer's *Le Prophète*, which Shaw wrote about as a music critic—and on another plane there is Wagner's *Ring of the Nibelung*, which Shaw saw as an allegorical dramatization of the historical process.

We have seen that according to Shaw his early reading of historical novels and plays prepared his mind for such writers as Macaulay, Hegel, and Marx: 'Macaulay did not repel me as a historian (of sorts).' The parenthetical qualification suggests Shaw's usual attitude towards Macaulay. It is difficult to say just how much of his work Shaw had read, but he was undoubtedly familiar with some of it. His treatment of late seventeenth-century English history in '*In Good King Charles's Golden Days*' strongly indicates an acquaintance with parts of Macaulay's *History of England*, and elsewhere Shaw refers to particular passages in the *History*. In his Preface to *The Doctor's Dilemma*, for example, he illustrated an argument by referring to Macaulay's description of Charles II's last illness (*CPP* iii. 261–2);[44] and in his Preface to *Cashel Byron's Profession* he used Macaulay's description of the flogging of Titus Oates as an example of the 'abominable vein of retaliatory violence all through the literature of the nineteenth century'.[45] He also reveals a knowledge of some of Macaulay's essays; in 1908 he suggested to Henry Arthur Jones, 'If you have Macaulay's essays handy, look at the beginning of the one on Robert Montgomery, which quotes a fable of Pilpay',[46] and in an 1897 review of a dramatization of *The Pilgrim's Progress*, he told his readers in an aside,

---

[43] Shaw, Preface to *Ellen Terry and Bernard Shaw*, p. xxxiv. Martin Meisel, in *Shaw and the Nineteenth-Century Theater* (Princeton: Princeton Univ. Press, 1963), suggests that the Dauphin's speech about his son, 'He that will be Louis the Eleventh', might be an allusion to Boucicault's play (352 n.)
[44] Cf. *Everybody's Political What's What?*, 214–15.
[45] Shaw, *Cashel Byron's Profession* (London: Constable, 1932), 15.
[46] Shaw to H. A. Jones, 8 Oct. 1908. MS in Harry Ransom Humanities Research Center, the University of Texas at Austin.

'[Y]es, thank you: I am quite familiar with Macaulay's patronizing prattle about The Pilgrim's Progress.'[47] Other references to Macaulay in Shaw's writing indicate knowledge of the poetry.

'Even if you find Macaulay so very readable that you waste on his history the time you should spend on more pregnant documents,' Shaw advised in 1917, 'the one thing that you do not learn from him is English history.' One of his defects is his narrowness of focus:

> Macaulay knew that modern democracy was making history a very important part of a common man's business; but he does not seem to have considered that our common democrats must, if they are to vote with any intelligence and exercise any real power, know not only the history of their own country, but that of all the other countries as well. Otherwise he would have bethought him that it is utterly impossible for common men to learn all these histories in such detail as he gives of his little parliamentary corner of the reigns of Charles II, James II, and William III.[48]

Most of Shaw's references to Macaulay are in this disparaging mode. He associated Macaulay with the self-satisfied Capitalism of the mid-nineteenth century, and regarded someone in the twentieth century who had not outgrown Macaulay as intellectually out of date. Being intellectually up to date is one of the subjects of *Man and Superman* (published in 1903), and one character who is distinctly behind the times is the young American husband of Violet, Hector Malone, *'whose culture is nothing but a state of saturation with our literary exports of thirty years ago'*. Hector attempts to educate England: *'When he finds people chattering harmlessly about Anatole France and Nietzsche, he devastates them with Matthew Arnold, the Autocrat of the Breakfast Table, and even Macaulay'* (*CPP* ii. 602). '[A]nd even Macaulay'—Hector is not just out of date but hopelessly, ludicrously out of date.

Sometimes Shaw uses Macaulay to represent the old dispensation, and Marx the new. In 1934 he explained this distinction in personal terms to St John Ervine: 'It is this Marxian debunking which has changed the mind of the whole world that you seem

---

[47] Shaw, *Our Theatres in the Nineties*, iii. 4.
[48] Shaw, 'Something Like a History of England', in *Pen Portraits and Reviews* (London: Constable, 1949), 89–90.

to have missed: you write like a man who had read no history after Macaulay, Taine, and Guizot. But I was completely Marxed....'[49] Or as he wrote to Winston Churchill in the same year: 'I read Karl Marx 14 years before Lenin did, and therewith got Macaulay out of my blood forever.' What one has to beware of in Macaulay in his Whiggery, Shaw told Churchill—'his idolatry of parliament, of public opinion, of "free" institutions, of an underlying divinity in the British character, and all the monstrous optimism which Karl Marx smashed like a conjuror kicking a very pretty lid off hell and shewing the horrible and grotesque realities beneath'.[50] In his Preface to *Farfetched Fables* in 1948–9 he argued that civil service examinations 'by elderly men of youths are at least thirty years out of date'; therefore the ill-instructed candidate will be successful, 'especially if he ranks [the views] of Karl Marx as blasphemous, and history as ending with Macaulay' (*CPP* vii. 414–15).

After his reading of *Das Kapital* in a French translation in 1883, Shaw tended for the rest of his life to call himself a Marxist, but his particular school of Marxism involved a rejection of most of Marx's analysis of society. We are not concerned here with the general issue of Shaw's response to Marx,[51] but Marx can be seen as a historian or as a philosopher-of-history in the nineteenth-century tradition, and Marx along with Bunyan (and others) could be a source for the linear, progressive element in Shaw's view of history. In any case, Marx is undoubtedly a source for his emphasis on historical epochs and economic factors in history. In the early Fabian years, Shaw recalled later, he and his colleagues promulgated a Marxist interpretation of historical development. 'As to history, we had a convenient stock of imposing generalizations about the evolution from slavery to serfdom and from

---

[49] Shaw to St J. Ervine, 23 Aug. 1934. MS in Harry Ransom Humanities Research Center, the University of Texas at Austin.
[50] Shaw to Winston Churchill, 8 May 1934, in Martin Gilbert, *Winston S. Churchill*, v, Companion Part 2, *The Wilderness Years 1929–1935* (London: Heinemann, 1981), 784–6. The context here is Shaw's response to Churchill's biography of his ancestor the Duke of Marlborough. Shaw's objection to the book—an outrageous provocation, in view of Churchill's desire to defend Marlborough against Macaulay—was that 'it is badly damaged in places by its Macaulayisms'. 'Now the stale parts of your book', Shaw wrote, 'are the Whig parts, the Macaulay parts.'
[51] There is a book which attempts to deal with this subject: Paul A. Hummert, *Bernard Shaw's Marxian Romance* (Lincoln: Univ. of Nebraska Press, 1973).

serfdom to free wage labor. . . . We gave lightning sketches of the development of the medieval craftsman into the manufacturer and finally into the factory hand.'[52] Shaw made it clear that he and the Fabian Society had left this crude Marxism behind; nevertheless, we may want to look for vestiges of it when we examine Shaw's mature view of the historical process.

Behind Marx lies Hegel, whose philosophy of history Shaw certainly knew something about—at least at second hand. Robert F. Whitman, who has studied the relationship between Shaw and Hegel, states that Shaw had read the Introduction to *The Philosophy of History* and 'perhaps some fragments elsewhere'. Whitman also notes that in 1887 Shaw read his friend Ernest Belfort Bax's *Handbook of the History of Philosophy*, which gave him an introduction to the philosophy of Hegel.[53] Given the general influence of Hegel in the late nineteenth century, it makes sense to see his teachings as part of Shaw's intellectual background, and we shall see that there is a strong Hegelian element in Shaw's historiography. One statement of Hegel's that we know Shaw was attracted to is in the Introduction to *The Philosophy of History*: '[W]hat experience and history teach is this—that peoples and governments never have learned anything from history, or acted on principles deduced from it.'[54] Shaw's version of this is pithier and more memorable: 'We learn from history that men never learn anything from history' (Preface to *Heartbreak House*, *CPP* v. 55).[55]

A philosopher of history whose work Shaw knew at first hand and well was Buckle. He read *The History of Civilization* at the outset of his career as a dramatist, and while there is much in Buckle that would not have appealed to him, the work as a whole impressed him greatly. He rejected Buckle's near-worship of Adam Smith and *laissez-faire* economics, and also the Voltairean rationalism that permeates *The History of Civilization*. On the

---

[52] Shaw, 'The Fabian Society: What It Has Done and How It Has Done It' (1892), in *Essays in Fabian Socialism* (London: Constable, 1949), 142–3.
[53] Robert F. Whitman, *Shaw and the Play of Ideas* (Ithaca: Cornell Univ. Press, 1977), 120, 152.
[54] G. W. F. Hegel, *The Philosophy of History*, tr. J. Sibree (New York: Dover, 1956), 6.
[55] Shaw cited this as his 'favorite Hegel epigram' in a letter to Janet Achurch, 8 Jan. 1895, quoted in Margot Peters, *Bernard Shaw and the Actresses* (Garden City, NY: Doubleday, 1980), 142. He also used it elsewhere in his writings.

Introduction

other hand, he was attracted to Buckle's belief in the saving power of sceptical independent enquiry, and his opposition to orthodoxy and censorship, and he would have admired Buckle's effort to see civilization as a whole, drawing upon not only political, constitutional history but also the arts, sciences, religion, and philosophy. Shaw would also have liked the attempt to find patterns underlying historical development—a Victorian taste which led him some years later to recommend Houston Stewart Chamberlain's *Foundations of the Nineteenth Century*: '[E]verybody capable of it should read it', he urged in the Preface to *Misalliance* (*CPP* iv. 132).

'Did you ever read Buckle's History of Civilization?' Shaw asked Allan Wade in 1908. 'If not, *do*.'[56] At the time that he himself read it he wrote a long letter to the man who had lent him the book, A. J. Marriott, explaining the reasons for his enthusiasm. 'I do not know how many years it is since you undertook to make me read Buckle's History of Civilization, and lent me your copy with that object,' Shaw's letter of 1894 began.

You must have despaired more than once of ever getting your three volumes back, much less inducing me to apply myself to them. But you will be glad to hear that I finished the last volume yesterday, having read every word of the three, notes and all, with the attention they deserve. And I assure you I am extremely obliged to you for making me do it. Out of the millions of books in the world, there are very few that make any permanent mark on the minds of those who read them. If I were asked to name some nineteenth century examples, I should certainly mention Marx and Buckle among the first.

The letter then goes on to compare Marx and Buckle, with emphasis on Buckle's 'intense belief in the value of doubt'.[57]

Shaw thought of Buckle and Marx together, as the historians of the economic basis.[58] In the Preface to *Back to Methuselah* (1921) he noted that Marx's theory of civilization 'had been promulgated already in Buckle's History of Civilization, a book as epoch-making in the minds of its readers as Das Kapital' (*CPP*

---

[56] Shaw to Allan Wade, 24 Nov. 1908, *Collected Letters 1898–1910*, ed. Dan H. Laurence (London: Max Reinhardt, 1972), 820.
[57] Shaw to A. J. Marriott, 28 Oct. 1894, *Collected Letters 1874–1897*, ed. Dan H. Laurence (London: Max Reinhardt, 1965), 456–8.
[58] See *The Religious Speeches of Bernard Shaw*, ed. Warren S. Smith (University Park: Pennsylvania State Univ. Press, 1963), 43, 67.

v. 314). And in the Appendix to *The Intelligent Woman's Guide to Socialism and Capitalism* (1928), in which he recommended further reading on these subjects, he said in his paragraph on Marx: 'His so-called Historic Materialism is easily vulnerable to criticism as a law of nature; but his postulate that human society does in fact evolve on its belly, as an army marches, and that its belly biases its brains, is a safe working one. Buckle's much less read History of Civilization, also a work of the mind changing sort, has the same thesis but a different moral: to wit, that progress depends on the critical people who do not believe everything they are told: that is, on scepticism.'[59]

This Appendix also recommends another historian who is of great importance to Shaw's work: Carlyle. Shaw includes Carlyle's *Past and Present* and 'Shooting Niagara', along with works by Ruskin, Morris, and Dickens, to represent 'the nineteenth-century poets and prophets who denounced the wickedness of our Capitalism exactly as the Hebrew prophets denounced the Capitalism of their time'.[60] In *Everybody's Political What's What?*, Shaw places Carlyle with Ruskin and Dickens as writers who rejected Capitalism without providing a political remedy. They 'would have none of Macaulay's cheerful meliorism and progress-boosting: they saw that Capitalism was the robber's road to ruin, and would not study its theory'. It took Marx, who was 'a trained Hegelian thinker', to propose an alternative political economy.[61] There is no direct evidence that Shaw had read widely or deeply in Carlyle's work, although his diaries record that in July 1888 he was reading *The French Revolution, Cromwell*, and *Frederick the Great*—'Then to the [British] Museum to read a bit of Carlyle until it was time to go home and go to the Opera', runs one entry.[62] The convincing in-direct-evidence of Shaw's interest in Carlyle is that there are many

[59] Shaw, *The Intelligent Woman's Guide to Socialism, Capitalism, Sovietism and Fascism* (London: Constable, 1949), 501. This work was first published in 1928 as *The Intelligent Woman's Guide to Socialism and Capitalism*; in 1937 the title was expanded to reflect the addition of two new chapters.
[60] *Intelligent Woman's Guide*, 503.
[61] *Everybody's Political What's What?*, 349–50.
[62] *Bernard Shaw: The Diaries*, ed. Stanley Weintraub (2 vols., University Park and London: The Pennsylvania State Univ. Press, 1986), i. 393–5. Elsewhere the diaries note that Shaw read Carlyle's essays on Richter and Chartism, and there is also a reference to the Cagliostro essay (i. 463, 582, 42).

striking similarities between Carlyle's writings and Shaw's, as will be clear when we come to look at *Heartbreak House*, for example. Shaw thought of Carlyle as a Victorian who 'never bowed the knee to Manchester',[63] and who 'called our boasted commercial prosperity shooting Niagara, and dismissed Cobdenist Free Trade as Godforsaken nonsense' (Preface to *Farfetched Fables*, *CPP* vii. 425–6); and Shaw's writing is filled with references to him. There is, for example, an unpublished letter to St John Ervine that reveals Shaw's liking for Carlyle's prose style: 'Browning', he wrote, 'captured me by the music of his verse, just as Carlyle did by the music of his prose. Both of them were reviled as discordant and uncouth because they left out the fashionable Tennysonian excess of sugar. I like it that way: I never took sugar in my tea.'[64] But Shaw's interest in Carlyle goes well beyond prose style. In my estimation, Carlyle is among the most important influences on Shaw's view of history, and my subsequent chapters will (I hope) demonstrate that a number of Shaw's plays can profitably be seen against the background of Carlyle's historical writing.

Shaw had read some of Nietzsche's work, and there is a discernible influence, but he chose to downplay the importance of Nietzsche in his intellectual development.[65] Ideas that critics attributed to Nietzsche, he said, came to him from quite another source: Buckle's travelling companion, who left him to die in Damascus in 1862—J. S. Stuart-Glennie, the author of books in the 1870s on patterns of historical development. In a letter to Archibald Henderson in 1905, Shaw described Stuart-Glennie as a 'Scotch historical philosopher' whose view of Christianity as a slave-morality is more sensible than Nietzsche's. 'Both views are obviously necessary to a grasp of the situation; but Nietzsche's is an impression, and Stuart Glennie's a piece of history.'[66] A year later Shaw took up this theme in his Preface to *Major Barbara*,

---

[63] Shaw to John Burns, 11 Sept. 1903, *Collected Letters 1898–1910*, 371.

[64] Shaw to St J. Ervine, 5 Feb. 1941. MS in Harry Ransom Humanities Research Center, the University of Texas at Austin. For a much earlier discussion of 'the freedom & directness as well as the force of [Carlyle's] mature style', see Shaw to [Pharall Smith], 11 Feb. 1890, *Collected Letters 1874–97*, 242.

[65] There is a chapter on Shaw's interest in Nietzsche in David S. Thatcher, *Nietzsche in England 1890–1914: The Growth of a Reputation* (Toronto: Univ. of Toronto Press, 1970), 175–217.

[66] Shaw to A. Henderson, 5 Sept. 1905, *Collected Letters 1898–1910*, 554.

where he insisted that he knew Stuart-Glennie's views on Christianity before he had ever heard of Nietzsche (*CPP* iii. 21). And in an obituary of Ibsen in the same year, he wrote that 'Nietzsche's theory of Christianity as a slave-morality is a mere clever guess compared to Stuart-Glennie's previous elaboration of it as the true basis of the philosophy of the history of civilisation since 600 B.C.'[67] Among Shaw's papers in the British Library is a printed piece by Stuart-Glennie on 'The Desirability of Treating History as a Science of Origins' and a typescript 'Plea for the Endowment of History as a Science'.[68] Shaw knew Stuart-Glennie personally, and in fact had lent him money.[69]

Shaw was also personally acquainted with other historical writers. In 1897 he arranged to read the newly completed *Devil's Disciple* to F. York Powell, who three years earlier had become Regius Professor of Modern History at Oxford University.[70] Then in 1940 Shaw corresponded with another Regius Professor of History (in this case at Cambridge), G. M. Trevelyan, in connection with '*In Good King Charles's Golden Days*'.[71] And one of his closest friends was Regius Professor of Greek at Oxford, Gilbert Murray, one of the great authorities on the civilization of ancient Greece.

Furthermore, Shaw became familiar with a number of historical works in the preparation of his own history plays. His main historical source for *The Devil's Disciple*, for example, was a work about General Burgoyne entitled *Political and Military Episodes in the Latter Half of the Eighteenth Century*, by Edward Barrington de Fonblanque. *Caesar and Cleopatra* owes much to *The History of Rome* by the German nineteenth-century historian Theodor Mommsen. *Saint Joan* owes even more to T. Douglas Murray's *Jeanne d'Arc*, which provides English translations of the transcripts of Joan's trial and the subsequent

[67] Shaw, 'Ibsen' (originally published in *The Clarion*, 1 June 1906), in *Shaw and Ibsen*, ed. J. L. Wisenthal (Toronto: Univ. of Toronto Press, 1979), 242.
[68] British Library Additional MS 50742.
[69] Shaw to J. S. Stuart-Glennie, 30 July 1908. MS in Harry Ransom Humanities Research Center, the University of Texas at Austin.
[70] Shaw to Charlotte Payne-Townshend, 28 Jan. 1897, *Collected Letters 1874–1897*, 723. For a brief, unsympathetic account of Powell's tenure as Regius Professor, see John Kenyon, *The History Men* (London: Weidenfeld and Nicolson, 1983), 177–8.
[71] See Shaw to G. M. Trevelyan, 20 Jan. 1940. MS in Harry Ransom Humanities Research Center, the University of Texas at Austin.

rehabilitation proceedings. In his late play, *The Six of Calais*, Shaw used Froissart's *Chronicle*, to which he had referred many years earlier in the Preface to *Major Barbara* (*CPP* iii. 27–8).

In all, Shaw wrote ten history plays—that is, if we define a history play narrowly as a work that is set in the historical past. In addition to *The Devil's Disciple*, *Caesar and Cleopatra*, *Saint Joan*, and *The Six of Calais*, there are Shaw's Napoleon play, *The Man of Destiny*; *The Glimpse of Reality*, a minor piece set in fifteenth-century Italy; his play about Shakespeare and Queen Elizabeth, *The Dark Lady of the Sonnets*; *Androcles and the Lion*, about Roman persecution of Christians; *Great Catherine*, a farcical treatment of the Russian empress; and Shaw's dramatic study of the Restoration period, '*In Good King Charles's Golden Days*'.

Then there are other historical subjects that Shaw considered writing plays about. He told Ellen Terry that he had an idea for a little play about King Pippin and his wife, 'with a lovely medieval French court for a stage setting',[72] and in 1903 he wrote to his German translator that he had 'thought of Cromwell, and may perhaps do him someday'.[73] A year later Cromwell was still on his mind: 'I want to write a one act piece called The Death of Cromwell,' he said to Harley Granville Barker. 'Cromwell & Napoleon would make a splendid historical program.'[74] In 1909 Shaw was reading an English translation of the Koran daily, as he prepared to do another history play: 'Mahomet's turn is coming: I shall write a play about him as a companion to Caesar, I think.'[75] In the same year, in showing a Parliamentary Committee on censorship of plays how the Lord Chamberlain's powers restricted the dramatist's choice of subjects, he explained why he had never fulfilled his long-standing desire to dramatize the life of Mahomet. The dramatic censorship raised problems which did not arise in the case of a work such as 'Carlyle's essay on the prophet', and the possibility that the Lord Chamberlain might

---

[72] Shaw to Ellen Terry, 26 Mar. 1896, *Collected Letters 1874–1897*, 617.
[73] Shaw to Siegfried Trebitsch, 7 Oct. 1903, *Collected Letters 1898–1910*, 376.
[74] Shaw to Granville Barker, 28 Aug. 1904, *Collected Letters 1898–1910*, 448. Dan H. Laurence reveals that as late as 1927 Shaw was still thinking about a Cromwell play (ibid. 446), although in 1920 Shaw wrote to his German translator that 'Cromwell has now been written by John Drinkwater: I have given him up' (Shaw to Siegfried Trebitsch, 15 Sept. 1920, *Collected Letters 1911–1925*, ed. Dan H. Laurence (London: Max Reinhardt, 1985), 688).
[75] Shaw to Lillah McCarthy, 19 Mar. 1909, *Collected Letters 1898–1910*, 837;

refuse to license a play about Mahomet had prevented Shaw from writing one. 'This restriction of the historical drama is an unmixed evil,' he declared. 'Great religious leaders are more interesting and more important subjects for the dramatist than great conquerors' (Preface to *The Shewing-Up of Blanco Posnet, CPP* iii. 713–14). As late as 1941, Shaw was telling St John Ervine that 'Luther is hopeless as a dramatic figure.... But I should write a play about Mahomet if I dared.'[76]

Out of the fifty-two plays that Shaw completed, then, ten are explicitly historical in subject matter. As I shall demonstrate in later chapters, many of his other plays have material that is in some sense historical. And his non-dramatic prose is full of references to historical forces, events, and characters. Shaw thinks historically, and it is natural to him to allude to the past in developing an argument. For an illustration of this tendency, we could go back to one of his early essays, in *Fabian Essays in Socialism* (1889). In 'The Transition to Social Democracy', Shaw began his consideration of this subject with a historical survey. 'Briefly, then, let us commence by glancing at the Middle Ages,' he said, and he provided an encapsulated economic history of England. In the other essay that he contributed to this volume, 'The Economic Basis of Socialism', he asserted that 'experience has lately convinced all economists that no exercise in abstract economics, however closely deduced, is to be trusted unless it can be experimentally verified by tracing its expression in history'.[77]

When Shaw wanted to explain Fascism in 1928, he referred his readers back to ancient Rome, seventeenth-century England, and nineteenth-century France. 'Julius Caesar, Cromwell, Napoleon and his nephew Louis Napoleon are the bygone Fascist leaders we talk most about' he wrote in *The Intelligent Woman's Guide to Socialism*. Similarly, in his 1945 Preface to *Geneva* he discussed

---

see also Shaw's letter to his wife in 1913, in which he reported that he was 'struggling with the temptation to begin another sketch—one which I planned long ago for Forbes Robertson—Mahomet in the slave market'. 'These thumbnail historical sketches amuse me', he told Charlotte, 'and the proceeds at the variety theatres & picture palaces will help to support your extravagance' (16 Aug. 1913, *Collected Letters 1911–1925*, 200).

[76] Shaw to St J. Ervine, 5 Feb. 1941. MS in Harry Ransom Humanities Research Center, the University of Texas at Austin.

[77] *Essays in Fabian Socialism*, 33–9, 24.

Hitler as a born leader by commenting that 'like Jack Cade, Wat Tyler, Essex under Elizabeth Tudor, Emmet under Dublin Castle, and Louis Napoleon under the Second Republic, [Hitler] imagined he had only to appear in the streets with a flag to be acclaimed and followed by the whole population' (*CPP* vii. 33). When in this same Preface he wanted to discuss England's victory over Hitler he did so by alluding to England's victories over Philip II of Spain, Louis XIV, Napoleon, and Kaiser Wilhelm (*CPP* vii. 17). In engaging in a controversy in 1940 about the trial of Oscar Wilde, Shaw insisted on the relevance of the trial of the Seven Bishops in 1688[78] (which is prominently and memorably described in Macaulay's *History of England*). His historical cast of mind is also evident in the way he chose to advance the cause of ex-servicemen after the First World War—in verse that decidedly says more for him as a historian than a poet:

> Justinian, in History's view
>   Your fame is not worth half a snowball
> Because, ungrateful monarch, you
>   Grudged Belisarius his obol.
>
> Again the veteran begs his bread
>   From you who swore he ne'er should rue it.
> For shame! It was for you he bled.
>   It is for you to see him through it.[79]

*

'To Bernard Shaw most of the past is simply a mess which ought to be swept away in the name of progress, hygiene, efficiency and what not.'[80] Shaw's work does provide some evidence to support this assertion of George Orwell's, and one could make a convincing argument to demonstrate that to Shaw, as to Henry Ford, history is bunk. Commenting on the school report card of a

---

[78] Letters to the editor of *The Times Literary Supplement*, in *Agitations: Letters to the Press 1875–1950*, ed. Dan H. Laurence and James Rambeau (New York: Frederick Ungar, 1985), 313–16.

[79] 'G.B.S., Lyrist', *The Living Age*, 8th series, 321 (Apr.–June 1924), 1017; the verse is reprinted from *The British Legion Album in Aid of Field-Marshall Earl Haig's Appeal for Ex-Service Men of All Ranks* (1924). In the 1920s, Shaw likened T. E. Lawrence to Belisarius, in trying to persuade the Prime Minister, Stanley Baldwin, to offer Lawrence a government pension; see *Collected Letters 1911–1925*, 831, 853, 910.

[80] George Orwell, *The Collected Essays, Journalism and Letters of George Orwell*, ed. Sonia Orwell and Ian Angus (London: Secker and Warburg, 1968), ii. 200.

friend's daughter, Shaw described history as 'an improper subject for young ladies if true, and a misleading one if false';[81] and there are passages in the plays that suggest history is false and misleading. In *The Devil's Disciple* when Burgoyne reveals that official bungling in London has cost England its American colonies, the simple-minded Major Swindon asks, 'What will History say?' —to which Burgoyne replies, 'History, sir, will tell lies, as usual' (Act III, *CPP* ii. 131). The Judge in *Geneva* refers to 'falsehoods called history' (Act III, *CPP* vii. 106),[82] and in *Saint Joan* the sophisticated Warwick explains to De Stogumber that 'It is only in history books and ballads that the enemy is always defeated' (Scene iv, *CPP* vi. 124).

To Shaw's Julius Caesar the past appears to be a mess which ought to be swept away in the name of progress. In Act II of *Caesar and Cleopatra* the tutor Theodotus announces with horror that the library of Alexandria is in flames. Caesar remains unmoved.

THEODOTUS. Without history, death will lay you beside your meanest soldier.
CAESAR. Death will do that in any case. I ask no better grave.
THEODOTUS. What is burning there is the memory of mankind.
CAESAR. A shameful memory. Let it burn.
THEODOTUS [*wildly*]. Will you destroy the past?
CAESAR. Ay, and build the future with its ruins.

(*CPP* ii. 219.)

It would be a naïve error to equate a character's views with the playwright's, but one could provide plenty of evidence that Shaw was much less interested in the past than in the future. In a speech in 1911 to the Heretics Society at Cambridge, for example, he advised his audience that 'They need not bother about the past.' 'Let the dead past bury the past,' he was quoted as saying in a report of the meeting. 'The concern of the Heretic was with the future, with the humanity that is to come.'[83] And when a French historian suggested Marat as a subject for a play in 1925, Shaw

---

[81] Shaw to Edith Livia Beatty, 7 Jan. 1900, *Collected Letters 1898–1910*, 134.
[82] Cf. a 1917 letter (not published) to the editor of *The Times* in which Shaw dismisses 'the heap of trumped-up special pleadings that are called history' (*Agitations*, 213).
[83] *The Religious Speeches of Bernard Shaw*, 36.

not only objected to Marat and Charlotte Corday as suitable figures for his drama, but he added: 'Besides, who cares for the French Revolution now? If I write a play about a revolution it will be the next one, not even the last one (the Russian).'[84] His next play was *The Apple Cart*, set not in the past but in the future.

In *Tragedy of an Elderly Gentleman* in *Back to Methuselah*, it is the slow-witted, outdated Elderly Gentleman who is interested in the past. It is he who foolishly boasts to the long-liver Zoo: 'Let me inform you that I can trace my family back for more than a thousand years, from the Eastern Empire to its ancient seat in these islands, to a time when two of my ancestors, Joyce Bolge and Hengist Horsa Bluebin, wrestled with one another for the prime ministership of the British Empire, and occupied that position successively with a glory of which we can in these degenerate days form but a faint conception' (Act I, *CPP* v. 507). Like Theodotus in *Caesar and Cleopatra*, he has a pedant's commitment to the retrieval of the past. In fact, he is the nearest thing to a professional historian in Shaw's plays. He is President of the Baghdad Historical Society, which 'has printed an editio princeps of the works of the father of history, Thucyderodotus Macollybuckle' (Act II, *CPP* v. 549–50). Although the Elderly Gentleman is allowed some words of wisdom in this play, his historical interests are not among them, and in the next play of *Back to Methuselah*, *As Far as Thought Can Reach*, we find the suggestion that the study of history is only an amusement for children, a diversion that they will outgrow when they reach maturity. One of the children in the play talks about 'the records which generations of children have rescued from the stupid neglect of the ancients' (*CPP* v. 585), and it is clear in the play that the neglectful Ancients are supposed to have our respect.

And yet both *Caesar and Cleopatra* and *Back to Methuselah*, which apparently invite us to reject history, are themselves very much concerned with historical issues. I want to discuss in a later chapter the way in which one can see *Back to Methuselah* as a play about time and historical process, but I should mention here one sentiment of the Elderly Gentleman's in *Tragedy of an Elderly Gentleman*. One of his dreams is 'to contemplate the ruins of London' (like the New Zealander in Macaulay's essay on

[84] Shaw to Augustin Hamon, 3 Dec. 1925, *Collected Letters 1911–1925*, 923.

Ranke's *History of the Popes*). 'Ruins!' exclaims Zoo. 'We do not tolerate ruins. Was London a place of any importance?'

THE ELDERLY GENTLEMAN [*amazed*]. What! London! It was the mightiest city of antiquity. [*Rhetorically*] Situate just where the Dover Road crosses the Thames, it—

ZOO [*curtly interrupting*]. There is nothing there now. Why should anybody pitch on such a spot to live? The nearest houses are at a place called Strand-on-the-Green: it is very old.

(Act I, *CPP* v. 530.)

What is being stressed here, as elsewhere in *Back to Methuselah*, is the importance of time and the reality of change. Whereas for Carlyle there is a still centre of enduring reality underlying historical change, Shaw's Life Force is dynamic, and the only reality is change. There are no permanent institutions: he asks his reader in *The Intelligent Woman's Guide* whether she notices 'that in these ceaseless activities which keep all of us fed and clothed and lodged, and some of us even pampered, NOTHING STAYS PUT? Human society is like a glacier: it looks like an immovable and eternal field of ice; but it is really flowing like a river.'[85] These capital letters reinforce an idea that is fundamental in Shaw's works. In political and social life there are no permanent institutions, and in intellectual life there are no permanent truths: 'All the assertions get disproved sooner or later', he declared in the Epistle Dedicatory to *Man and Superman* (*CPP* ii. 527). If change is the constant and fundamental reality of our experience, then one cannot dismiss history so easily.

For Shaw, change is not only the fundamental reality but it is the main ground for optimism. The changes in *Tragedy of an Elderly Gentleman* are painful for the Elderly Gentleman, but the audience is to accept them as inevitable and desirable. Lilith declares at the end of the last play, *As Far as Thought Can Reach*, 'I say, let them [men and women] dread, of all things, stagnation' (*CPP* v. 630), and Shaw always comes back to the conflict between static death and dynamic life. The fact that no institution is permanent makes it possible to hope for, and work for, the supersession of the present forms by something better. Part of the value of studying history, therefore, is to recognize that no

---

[85] *Intelligent Woman's Guide*, 308.

institution is permanent. In arguing that we must scrap the system of political parties, Shaw says that most people regard this system as an inevitable part of public affairs, part of an unchanging human nature. Thus 'proposals to drop The Party System are usually dismissed without examination as Utopian attempts to get rid of political parties'. It must be explained to people, therefore, 'that The Party System, forced on William III at the end of the seventeenth century to secure the support of Parliament for his war against Louis XIV; vigorously repudiated by Queen Anne; but finally established during the eighteenth century as our normal constitutional method of parliamentary Government, means simply the practice of selecting the members of the Government from one party only'.[86] If people can be made to see the party system in a historical context, then they will realize that it is merely a temporary historical phenomenon that can be abolished. Shaw's view is pretty much the obverse of Edmund Burke's: for Shaw, an institution that has developed historically has no special claim to continued existence. All institutions are thus merely temporary. 'Capitalism,' Shaw explained in *The Intelligent Woman's Guide*, 'though it has destroyed many ancient civilizations, and may destroy ours if we are not careful, is with us quite a recent heresy, hardly two hundred years old at its worst.'[87]

The historical perspective gives hope for the future, and it also gives the knowledge that is necessary for any understanding of the future. Shaw would have agreed with Hobbes's view: 'No man can have in his mind a conception of the future, for the future is not yet. But of our conceptions of the past, we make a future.'[88] This is precisely the method of *Back to Methuselah*, and it is what Shaw advises his readers in *The Intelligent Woman's Guide*: 'Until we know what has happened to the Changed we shall not understand what is going to happen to the Not Yet Changed.'[89]

The historical perspective also gives the knowledge that is necessary for any understanding of one's present society. 'I can

---

[86] Shaw, 'Fabian Essays Forty Years Later', in *Essays in Fabian Socialism*, 310.
[87] *Intelligent Woman's Guide*, 128.
[88] Thomas Hobbes, *The Elements of Law: Natural and Politic*, quoted in Philip Rosenberg, *The Seventh Hero: Thomas Carlyle and the Theory of Radical Activism* (Cambridge, Mass.: Harvard Univ. Press, 1974), 64.
[89] *Intelligent Woman's Guide*, 310.

... assure you that the way to understand the changes that are going on is to understand the changes that have gone before, and warn you that many women have spoilt their whole lives and misled their children disastrously by not understanding them.'[90] It is not possible for us to grasp contemporary events in the way that we can grasp events of the past. 'No epoch is intelligible until it is completed and can be seen in the distance as a whole, like a mountain. The victorious combatants in the battle of Hastings did not know that they were inaugurating feudalism for four centuries, nor the Red Roses on Bosworth Field and the Ironsides at Naseby know that they were exchanging it for Whig plutocracy.' This is why it is essential for political leaders to have a knowledge and understanding of past events. 'Statesmen who know no past history are dangerous because contemporary history cannot be ascertained' (Preface to *Geneva, CPP* vii. 18).

The main purpose of Shaw's book *Everybody's Political What's What?* is to state what a citizen must know in order to be eligible for participation in public affairs. On the final page of the book Shaw speaks of the ignorance from which contemporary politicians suffer: 'Macaulay's history of England and the Communist Manifesto of Marx and Engels are not infallible scriptures; but persons who have never read them nor comprehended the change in historical outlook from one to the other should be eligible for the Foreign Office, or indeed any Downing Street office, only as porters or housemaids. Yet we never dream of asking whether a Secretary of State has ever heard of Macaulay or Marx, nor even whether he can read the alphabet.'[91] Politicians' ignorance of history is a recurrent theme in Shaw's writing. In the Preface to *The Shewing-Up of Blanco Posnet*, for example, he accused the report of the Parliamentary Committee on stage censorship of proposing a new Star Chamber, and he commented: 'Now I have no guarantee that any member of the majority of the Joint Select Committee ever heard of the Star Chamber or of Archbishop Laud. . . . Nothing is more alarming than the ignorance of our public men of the commonplaces of our history, and their consequent readiness to repeat experiments which have in the past produced national catastrophes' (*CPP* iii.

---

[90] *Intelligent Woman's Guide*, 309.
[91] *Everybody's Political What's What?*, 366.

745–6). One such public man in Shaw's plays is the fatuous Ambrose Badger Bluebin, Prime Minister of the British Islands in *Tragedy of an Elderly Gentleman* in *Back to Methuselah*. When his long-lived guide tells of the extinction of western civilization in the period following the First World War, he more or less boasts, 'I dont know any history: a modern Prime Minister has something better to do than sit reading books' (Act II, *CPP* v. 541–2).

*

In 1917 Shaw, in reviewing G. K. Chesterton's *A Short History of England*, said that the book 'raise[s] hopes that the next generation may learn something of what it needs to know about the history of its own country'.[92] A year later he wrote 'A Glimpse of the Domesticity of Franklyn Barnabas', originally intended to be part of *The Gospel of the Brothers Barnabas* in *Back to Methuselah*. One of the characters, Immenso Champernoon, is based on Chesterton, and one passage of dialogue has to do with attitudes towards history.

IMMENSO. . . . But I still defend the past. My roots are in the past: so are yours.

MRS ETTEEN. My roots are in the past: my hopes are in the future.

IMMENSO. The past is as much a part of eternity as the future. Beware of ingratitude to the past. What is gratitude? The cynic says, a sense of favors to come. Many people would say that we cannot be grateful to the past because we have nothing to expect from it. But if we had no memory of favors from the past we could have no faith in favors from the future.

(*CPP* v. 680.)

Of course, Immenso's views here are not necessarily Shaw's, any more than Caesar's views on the burning of the library at Alexandria are Shaw's. We can conclude, however, that Shaw considered a knowledge of history to be fundamental for anyone who wishes to think about or participate in public affairs. In fact his dismissal of the past and his assertion of its importance are not altogether contradictory: history for him is concerned with the past in relation to the present, and the present in relation to the past. An isolated concern with the past for its own sake is sterile,

---

92 Shaw, 'Something Like a History of England', in *Pen Portraits and Reviews*, 89.

but one cannot understand the present without a historical context. Past and present are of a piece, and a knowledge of both is necessary to enable one to think intelligently about the future. Shaw's own view *is* expressed, I believe, by a character in his late play about the future, *Farfetched Fables*: 'As I can neither witness the past nor foresee the future I must take such history as there is as part of my framework of thought. Without such a framework I cannot think any more than a carpenter can cut wood without a saw' (Sixth and Last Fable, *CPP* vii. 461). In spite of the dismissals of history in parts of Shaw's writing, he held the study of history to be essential.

iii. Shaw's Historiography

Shaw usually gives the impression that existing historical work is inadequate. In his single address in the United States, at the Metropolitan Opera House in New York City in April 1933, he said to his audience: 'You know, if you study American history —not the old history books; for almost all American histories, until very lately, were mere dustbins of the most mendacious vulgar journalism—but the real history of America, you will be ashamed of it because the real history of all mankind is shameful.'[93] He did not disclose which recent historians he regarded as reliable; as usual his emphasis was on the defects of historians. We saw in the previous section that Shaw was not impressed with the quality of the historical education that he himself had received in school; and he deplored the historical teaching available in educational institutions generally. A character in *Buoyant Billions* claims that at Oxford you cannot hope for a degree '[u]nless you are a hundred years behind hand in science and seven hundred in history' (Act II, *CPP* vii. 331), and Shaw himself claimed in *Everybody's Political What's What?* that the public schools teach 'fabulous history'—for example that 'the battles of Trafalgar and Waterloo, which substituted Louis XVIII for Napoleon as a fitter ruler of France, were triumphs of civilization and British good sense'.[94] Educated people have been

---

[93] Shaw, *The Political Madhouse in America and Nearer Home* (London: Constable, 1933), 32.
[94] *Everybody's Political What's What?*, 147.

badly educated, and history is known to conventional people, 'when it is known at all, as a string of battles in which their side has been victorious'.[95]

What approaches to history would Shaw regard as acceptable and desirable? This is not an easy question to answer simply, because Shaw's method of argument is to state one side of a case strongly in one place, and then another side of the case just as strongly elsewhere.

There are passages in Shaw's writing that suggest a Marxist view of history, in the loose sense that economic forces are seen as fundamental. This is the viewpoint that predominates in *The Intelligent Woman's Guide to Socialism*. In a chapter entitled 'Money', for example, Shaw argues that those who understand the issues are far more concerned with Henry VIII's debasement of the coinage than with the fact that he married six wives or allowed the nobles to plunder the Church. Another chapter asserts that the First World War was caused not by human wickedness or even human intention, but rather by European competition for African markets.[96] *Everybody's Political What's What?* carries this argument further. Wars occur, we discover in a chapter headed 'The Economic Man', when the interest rate has fallen from 5 per cent to 2.5 per cent. 'The truth is that the need of Capital for death and destruction sets in motion the human force of natural pugnacity.'[97] Much of the argument of both of these volumes is summed up in the statement that 'in the long run money will take us anywhere, and lack of it will tame the haughtiest peer'.[98]

Shaw sometimes likes to argue, too, that writers and thinkers play little part in the processes of history. '[C]ivilized life is changed much more sensationally by inventions than by books.... Shakespear made comparatively no social changes: Watt and Stephenson made the industrial revolution.'[99] One of the most emphatic expressions of this point of view is to be found in the Preface to *Farfetched Fables*, written when Shaw was in his nineties.

[95] *Intelligent Woman's Guide*, 479.
[96] *Intelligent Woman's Guide*, 254, 152–7.
[97] *Everybody's Political What's What?*, 143–5.
[98] *Everybody's Political What's What?*, 58.
[99] *Everybody's Political What's What?*, 99.

Homilies cut no ice in administrative councils: the literary talent and pulpit eloquence that has always been calling for a better world has never succeeded, though it has stolen credit for many changes forced on it by circumstances and natural selection. The satirical humor of Aristophanes, the wisecracks of Confucius, the precepts of the Buddha, the parables of Jesus, the theses of Luther, the *jeux d'esprit* of Erasmus and Montaigne, the Utopias of More and Fourier and Wells, the allegories of Voltaire, Rousseau, and Bunyan, the polemics of Leibniz and Spinoza, the poems of Goethe, Shelley, and Byron, the manifesto of Marx and Engels, Mozart's Magic Flute and Beethoven's Ode to Joy, with the music dramas of Wagner, to say nothing of living seers of visions and dreamers of dreams: none of these esthetic feats have made Reformations or Revolutions.

(*CPP* vii. 427.)

The play that (on the surface, at any rate) comes the closest to this attitude is *Major Barbara*. In the Preface to this play we have another list of thinkers who have failed to change the world, and a discussion of the French Revolution which rejects the view that it was the work of Voltaire, Rousseau, and the Encyclopaedists. Rather, it was the work of 'men who had observed that virtuous indignation, caustic criticism, conclusive argument and instructive pamphleteering, even when done by the most earnest and witty literary geniuses, were as useless as praying, things going steadily from bad to worse whilst the Social Contract and the pamphlets of Voltaire were at the height of their vogue'. Even 'perfectly respectable citizens and earnest philanthropists' supported the September Massacres, because they realized there was no other way to prevent the aristocracy from regaining power (*CPP* iii. 38–9).

But in *Major Barbara* itself the Professor of Greek brings his lucidity of mind to the munitions works; the thinker has an essential part to play in the historical process. (I shall return to *Major Barbara* in a later chapter—I consider it to be one of Shaw's main history plays.) As for the French Revolution, here is another passage from *Everybody's Political What's What?*: 'Voltaire, Diderot and Rousseau made Robespierre and Napoleon possible. Lassalle and Marx, Engels and Richard Wagner, made Hitler and Mussolini possible as well as Lenin, Stalin, and Ataturk. Carlyle and Ruskin, Wells and Shaw, Aldous Huxley and Joad, are making possible the devil knows who in

England: probably someone of whom these sages would vehemently disapprove.'[100]

Now, one could argue that Shaw simply contradicts himself, and that his thinking is muddled and not worth paying much attention to. I believe on the contrary that Shaw's works as a whole, if we allow for deliberate over-emphasis in particular assertions, advance a consistent and intelligible argument. That argument is that thinkers are a necessary condition of historical development, but not a sufficient condition. Historical change occurs when thought is incarnated in men (and women) of action. Thought is impotent when separated from the sphere of action, but thought has primacy. It is the preliminary condition for historical change. Shaw's plays are closer to Hegel's view of history than to Marx's, in that they present ideas as anterior to material conditions. In *In the Beginning*, the first of the *Back to Methuselah* plays, the Serpent explains to Eve that 'imagination is the beginning of creation. You imagine what you desire; you will what you imagine; and at last you create what you will' (*CPP* v. 348). *Back to Methuselah* reveals the primacy of thought, of imagination, of belief, of religion, in historical development. Adam and Eve imagine birth and death; then birth and death come into being. The Brothers Barnabas imagine an extension of the human lifespan; then it comes into being. Ideas rather than economic forces are the primary cause, as they are in *Saint Joan*, where it is Joan's vision and not any economic factor that is primarily responsible for the expulsion of the English from France. The Reformation begins in the minds of prophetic people like Joan, and not in an economic imperative. *Saint Joan* is a dramatization of the emergence of a new era in history as a result of a new conception of the world.

The Preface to *Saint Joan* reviews the defects of previous literary attempts to deal with Joan's history. When Shaw comes to the late Victorian man of letters, Andrew Lang, he comments that 'like Walter Scott, he enjoyed medieval history as a string of Border romances rather than as the record of a high European civilization based on a catholic faith' (*CPP* vi. 44). This criticism suggests Shaw's own way of looking at historical epochs as the embodiment of an idea or a religion. His dramatic practice is in

[100] *Everybody's Political What's What?*, 310.

accord with the Hegelian arguments of R. G. Collingwood's *The Idea of History*. Collingwood accepts Hegel's position 'that there is no history except the history of human life, and that, not merely as life, but as rational life, the life of thinking beings', and says it follows from this that 'all history is the history of thought'. 'Here again Hegel was certainly right; it is not knowing what people did but understanding what they thought that is the proper definition of the historian's task.' Elsewhere in his study, under the heading 'History as Knowledge of Mind', Collingwood elaborates on this:

> History, then, is not, as it has so often been mis-described, a story of successive events or an account of change. Unlike the natural scientist, the historian is not concerned with events as such at all. He is only concerned with those events which are the outward expression of thoughts, and is only concerned with these in so far as they express thoughts. At bottom, he is concerned with thoughts alone; with their outward expression in events he is concerned only by the way, in so far as these reveal to him the thoughts of which he is in search.[101]

This nicely describes the historiography of Shaw's history plays, and it indicates why Shaw makes the 'sacrifice of verisimilitude' that he discusses in the Preface to *Saint Joan*. '[I]t is the business of the stage to make its figures more intelligible to themselves than they would be in real life; for by no other means can they be made intelligible to the audience,' Shaw said. Therefore he has made his Cauchon, Lemaître, and Warwick more articulate and intellectually self-conscious than their historical originals would have been. Shaw's claim is that 'as far as I can gather from the available documentation, and from such powers of divination as I possess, the things I represent these three exponents of the drama as saying are the things they actually would have said if they had known

---

[101] Collingwood, *The Idea of History*, 115, 217. Shaw would have found similar passages in his reading of Buckle's *History of Civilization*, e.g. (in a discussion of the causes of the French Revolution): 'That to which attention is usually drawn by the compilers of history is, not the change, but is merely the external result which follows the change. The real history of the human race is the history of tendencies which are perceived by the mind, and not of events which are discerned by the senses. It is on this account that no historical epoch will ever admit of that chronological precision familiar to antiquaries and genealogists. The death of a prince, the loss of a battle, and the change of a dynasty, are matters which fall entirely within the province of the senses; and the moment in which they happen can be recorded by the most ordinary observers. But those great intellectual revolutions upon which all other revolutions are based, cannot be measured by so simple a standard' (ii. 324–5).

what they were really doing. And beyond this neither drama nor history can go in my hands' (*CPP* vi. 73–4). In enabling his characters to express what they would have articulated 'if they had known what they were really doing', Shaw is dramatizing history as the history of thought, of ideas.

One good way to look at Shaw's dramatic historiography is to set some of his history plays next to their Victorian antecedents, as Martin Meisel does in his *Shaw and the Nineteenth-Century Theater*. One can compare Shaw's *Saint Joan*, for example, with Tennyson's *Becket*.[102] In both plays we have a medieval setting, a conflict between Church and State, and the martyrdom of a saint. The rejection of Joan at the end of the Cathedral scene in Shaw's play could well owe something to Tennyson's portrayal of the rejection of Becket in Act II, scene ii of his play. But in Tennyson's play the issues are personal ones, while in Shaw's the conflict is between opposing ideas. Meisel provides the best statement of Shaw's historiography in the history plays: 'For Shaw, the "essential truth" of any historical conflict lay in the ideas (and the institutions insofar as they embodied the ideas) at stake in the conflict. Consequently, Shaw's history-makers are the men and women who embody passionate ideas, dramatically articulating and expounding themselves. . . . Shaw as historian belonged very much to the idealist schools of the nineteenth century; for he presented ideas, embodied in men, as the realities of history, and will, not accident, as its driving energy.'[103] Whereas in nineteenth-century history plays (Scribe's *The Glass of Water* is a particularly striking example), love affairs and little accidents determine the direction of history, in Shaw's history plays the motive force is the human will, which gives expression to conceptions created by the human mind.

*

Shaw's dramatic treatment of historical sources reflects this conception of history as essentially the history of ideas: what people thought as opposed to what they merely did. He has been quoted as saying late in his life that his history plays were 'easy to

---

[102] Meisel discusses *Becket*, and he observes that 'the issues suggested in Tennyson's play are very close to the historical issues expounded in *Saint Joan*' (*Shaw and the Nineteenth-Century Theater*, 353).
[103] Meisel, *Shaw and the Nineteenth-Century Theater*, 374–5. Meisel's whole chapter on 'Historical Drama' is illuminating.

write because the facts are a gift: all that has to be done is to supply the ideas and the people personifying the ideas'.[104] Let us look first at the way in which Shaw drew on the historical facts which were a gift, and then at the secondary role these facts play in his history plays.

One of the speakers in Wilde's 'The Decay of Lying' (1889) complains that the writing of history has degenerated since the time of Carlyle, whose *French Revolution* 'is one of the most fascinating historical novels ever written', and in whose works 'facts are either kept in their proper subordinate position, or else entirely excluded on the general ground of dulness'. A manifestation of the decay of lying is the phenomenon that 'Now everything is changed. Facts are not merely finding a footing-place in history, but they are usurping the domain of Fancy, and have invaded the kingdom of Romance. Their chilling touch is over everything. They are vulgarising mankind.'[105] The name that is most commonly associated with the dominion of fact in nineteenth-century historiography is that of the German historian Leopold von Ranke (1795–1886), who announced in the 1830s that whereas history has had assigned to it the task of judging the past, of instructing the present for the benefit of future ages, his purpose was 'merely to show how things actually were'. He later claimed that he had 'resolved to avoid all invention and imagination' in his works 'and to stick to facts'.[106]

For the most part Shaw rejected the importance of fact in history and in historical drama, but there are places in his writing where he sounds more Rankean than usual. In the conclusion to *Everybody's Political What's What?*, he acknowledged that 'Though history is adulterated with lies and wishful guesses, yet it sifts and sheds them, leaving finally great blocks of facts.'[107] The historical dramatist works with these great blocks of facts. When an interviewer asked Shaw in 1938 whether a dramatist who is

[104] Stephen Winsten, *Days with Bernard Shaw* (London: Readers Union, Hutchinson, 1951), 235.
[105] Oscar Wilde, 'The Decay of Lying', *The Works of Oscar Wilde*, ed. G. F. Maine (London and Glasgow: Collins, 1948), 919.
[106] Quoted in Gooch, *History and Historians in the Nineteenth Century*, 74. For a discussion of Ranke's pre-eminence in nineteenth-century historiography, see E. H. Carr, *What Is History?*, 3–5.
[107] *Everybody's Political What's What?*, 366.

writing a historical play should be allowed to 'clothe his characters in garbs of romance', he replied: 'If the characters are clothed in the garb of romance, as you so romantically put it, they are not historical. No historical character is worth dramatizing at all unless the truth about him or her is far more interesting than any romancing. . . . Shakespeare always stuck close to the chronicles in his histories. And they survive, whilst hundreds of pseudo-historical plays have perished.'[108] At the end of 1897 Shaw was reading up the subject of Peter the Great (while visiting at the home of Macaulay's nephew and biographer Sir George Otto Trevelyan), and a week later his review of Laurence Irving's play *Peter the Great* appeared in the *Saturday Review*. Shaw used his reading in order to demonstrate the historical inaccuracies of Irving's play, and it is revealing that he invoked historical accuracy as a criterion in judging the play.[109]

He also applied this criterion to his own history plays. Take *The Devil's Disciple*, for example. The Harry Ransom Humanities Research Center at the University of Texas at Austin has a book request form from the British Museum which is covered with Shaw's detailed notes on Burgoyne, and when Shaw was challenged on the accuracy of his Burgoyne he defended the play as 'authentic history' in an article entitled 'Trials of a Military Dramatist' (See *CPP* ii. 151–6). When the play was published in 1901 as one of the *Three Plays for Puritans*, Shaw included 'Notes to The Devil's Disciple', which discussed the historical Burgoyne (with a brief note too on the chaplain Brudenell).

The same volume of plays contains *Caesar and Cleopatra*, with another set of Notes. For the copyright performance of the play in 1899, Shaw prepared a programme note which paraded his sources as impressively as possible, as a parody of scholarship:

> The play follows history as closely as stage exigencies permit. Critics should consult Manetho and the Egyptian Monuments, Herodotus, Diodorus, Strabo (Book 17), Plutarch, Pomponius Mela, Pliny, Tacitus, Appian of Alexandria, and, perhaps, Ammianus Marcellinus.
>
> Ordinary spectators, if unfamiliar with the ancient tongues, may refer to Mommsen, Warde-Fowler, Mr. St. George Stock's Introduction to

---

[108] 'The Theatre To-day and Yesterday', *Manchester Evening News*, 6 Dec. 1938, 6.
[109] Shaw to Ellen Terry, 31 Dec. 1897, *Collected Letters 1874–1897*, 838–9; *Our Theatres in the Nineties*, iii. 283–8.

the 1898 Clarendon Press edition of Caesar's Gallic Wars, and Murray's Handbook for Egypt. (*CPP* ii. 306.)[110]

It is fairly certain that Shaw had not himself consulted most of these authorities. Gale K. Larson, in an article on the composition of *Caesar and Cleopatra*, suggests that Shaw's recommended readings 'are merely lifted from bibliographical listings contained in two sources'—St George Stock's 'Introduction' and Murray's *Handbook*.[111]

There is, however, at least one work that he did consult closely in writing *Caesar and Cleopatra*: Theodor Mommsen's *History of Rome*, which he read (in part, at any rate), in an English translation by William P. Dickson. '[M]y Caesar is Mommsen's Caesar dramatized', Shaw told his German translator in 1906,[112] and in 1918 he wrote to Hesketh Pearson:

[*Caesar and Cleopatra*] is what Shakespeare called a history: that is, a chronicle play; and I took the chronicle without alteration from Mommsen. I read a lot of other stuff, from Plutarch, who hated Caesar, to Warde-Fowler; but I found that Mommsen had conceived Caesar as I wished to present him, and that he told the story of the visit to Egypt like a man who believed in it, which many historians dont. I stuck nearly as closely to him as Shakespeare did to Plutarch or Holinshed.[113]

Shaw was amused by the obtuseness of most of the critics, who (he claimed) did not see how closely he had followed his sources. Only two critics 'knew that what they were looking at was a chapter of Mommsen and a page of Plutarch furnished with scenery and dialogue, and that a boy brought to see the play could pass an examination next day on the Alexandrian expedition without losing a mark' ('*Caesar and Cleopatra*, by the Author of the Play', *CPP* ii. 312). Shaw thought that the reviews of the

---

[110] The Harry Ransom Humanities Research Center, the University of Texas at Austin has a copy of this programme note with autograph revisions by Shaw for the 1906 Forbes-Robertson production in New York. Shaw has altered the beginning of the programme note to read: 'The play follows history so closely that critics are respectfully recommended not to reject any incident as fictitious before consulting Manetho', etc.

[111] Gale K. Larson, '*Caesar and Cleopatra*: The Making of a History Play', *Shaw Review*, 14 (May 1971), 83. The whole article (pp. 73–89) is a valuable discussion of Shaw's use of sources and treatment of history in the play.

[112] Shaw to Siegfried Trebitsch, 13 Apr. 1906, *Collected Letters 1898–1910*, 616.

[113] Hesketh Pearson, *G.B.S.: A Full Length Portrait* (New York and London: Harper and Brothers, 1942), 187.

American production in 1906 displayed a particular ignorance of what was history and what was Shavian invention. '[E]verything in *Caesar and Cleopatra*, which is simply dramatised Mommsen or transcribed Plutarch, has been pooh poohed as fantastic modern stuff of my own, whilst the few modern topical allusions I have indulged in, including the quotation from Beaconsfield on Cyprus, have passed unchallenged as grave Roman history' ('Bernard Shaw and the Heroic Actor', *CPP* ii. 308–9).

Shaw, then, can be insistent about the historical accuracy of *Caesar and Cleopatra*, and there is evidence that he went to some trouble to get the historical details right. The Harry Ransom Humanities Research Center has manuscript notes that Shaw took from Mommsen's *History* in 1898, and also extracts from a French translation of the life of Julius Caesar by Suetonius, in Charlotte Shaw's handwriting—a passage about the Alexandrian episode.[114] Shaw consulted with his friend Gilbert Murray while the play was in manuscript, and wrote to him in 1900, 'I have carefully considered your comments on my history, and have modified accordingly.'[115] He made a number of minor changes when publishing the play in 1901 in *Three Plays for Puritans*, in order to improve its historical accuracy.[116]

In 1912, when Shaw was completing *Androcles and the Lion*, he once again called upon Murray for expert advice on details. He asked Murray 'to go over it and correct any howlers, also to give me in charity some Roman names that are not hackneyed to death: I can think of nothing but Metellus &c &c; and I dont know which emperor my Caesar should be—not Nero if I can possibly help it'.[117] Murray commented favourably on the play, and suggested, 'The names seem to me quite good, except that it ought to be Megaera, not Mag.'[118] By the time the play reached the rough proof stage, Shaw had made the change. The extent of

[114] For Shaw's use of Mommsen in *Caesar and Cleopatra* see Stanley Weintraub, *The Unexpected Shaw* (New York: Frederick Ungar, 1982), 111–23.
[115] Shaw to Gilbert Murray, 28 July 1900, *Collected Letters 1898–1910*, 179–80.
[116] I have written about Shaw's concern for historical accuracy in *Caesar and Cleopatra* in my introduction to the *Man of Destiny* and *Caesar and Cleopatra* volume of *Early Texts: Play Manuscripts in Facsimile*, ed. Dan H. Laurence (New York and London: Garland, 1981), pp. xvii–xviii.
[117] Shaw to Gilbert Murray, 3 Feb. 1912, *Collected Letters 1911–1925*, 74.
[118] G. Murray to Shaw, 23 July and 30 July [1912]. MSS in Harry Ransom Humanities Research Center, the University of Texas at Austin.

Shaw's reliance on Murray is humorously indicated in a comment Shaw made to him in 1928: 'My reputation as an expert in early Christianity (founded on Androcles) is entirely due to your secret instructions.'[119]

In preparing for the London production of *Saint Joan* in 1924, Shaw consulted with another of his friends, Sydney Cockerell, director of the Fitzwilliam Museum in Cambridge and formerly an associate of William Morris. Shaw corresponded with Cockerell about Warwick as a crusader, and about heraldic and ecclesiastical details.[120] It was Cockerell who had given him a copy of T. Douglas Murray's *Jeanne d'Arc*, which provided most of the historical details of the play; 'My Plutarch', Shaw said, 'was the report of the trial and the rehabilitation: contemporary and largely verbatim. I took particular care not to read a word of anything else until the play was finished.'[121] Shaw claimed to have followed the record closely: 'My account is essentially correct and historical', he insisted in 1948 ('Shaw's Saint Joan', *CPP* vi. 243), and in a 1924 interview he is quoted as making the Rankean claim that 'I have merely written a play based upon the facts as they exist.'[122] For the first London production in 1924 he wrote a programme note that described how far the play departs from historical facts. Not very far at all, Shaw asserted: 'It does not depart from ascertainable historical truth in any essential particular.' It is just that 'historical facts cannot be put on the stage exactly as they occurred, because they will not fit into its limits of time and space'; and Shaw gives examples of necessary condensation. One such example is the trial, which has been reduced from several days to forty minutes, 'but nothing essential is misrepresented; and nothing is omitted except the adjournments and matters irrelevant to the final issue'. As in the case of *The Devil's Disciple*, the play is closer to the historical record than it may sound in places: 'Several of the speeches and sallies in the play, especially those of Joan, are historical; and some of them may possibly sound like modern jokes; for instance, her use of the

[119] Shaw to G. Murray, 23 May 1928. MS in British Theatre Association Library.
[120] Shaw to S. Cockerell, 27 Feb. 1924, *Collected Letters 1911–1925*, 866–8.
[121] Shaw to John Middleton Murry, 1 May 1924, *Collected Letters 1911–1925*, 875.
[122] Walter Tittle, 'Mr. Bernard Shaw Talks about St. Joan', *The Outlook*, 25 June 1924, reprinted in Stanley Weintraub, ed., *'Saint Joan' Fifty Years After* (Baton Rouge: Louisiana State Univ. Press, 1973), 14.

word *godons* (God damns) to denote English soldiers' ('Note by the Author', *CPP* vi. 212–15).

Saint Joan, Shaw told Archibald Henderson, 'has been, of all my plays, the easiest one to write—because such incidents as those of the trial can be transcribed almost literally from the original documents'.[123] And in a letter to Cockerell about the play he made the same point in greater detail: 'I am not inordinately proud of it. You see, it was very easy to write: the materials were there, and even the historical manufacture had been worked over by so many hands that I am only the author in the sense that Michael Angelo was the architect of St Peter's: Ruskin and Morris and all the painters were on the job before me: I have had only to pull it together and fit it in.'[124] One might regard the Epilogue as the play's obvious departure from the historical record, but even the Epilogue can be seen as a very free dramatic treatment of the 1456 Rehabilitation proceedings, which are included in Murray's book.

For '*In Good King Charles's Golden Days*', Shaw once again had the advantage of a consultant, this time George Macaulay Trevelyan. There is a postcard from Shaw to Trevelyan in the Harry Ransom Humanities Research Center that indicates the kind of assistance that Trevelyan provided with points of historical detail.

The play was read by two astronomers (Jeans + Eddington), by the omniscient Webbs, and the entire personel [*sic*] of the Malvern Festival. They all swallowed the Calvin blunder without a grimace. I was quite right about it myself; but most unfortunately instead of trusting to my historical instinct (the true method) I foolishly looked up the date in the Enc. Br. and of course was set wrong, misreading the century because my head was full of the seventeenth. I dont know why I made Charles say he couldnt remember his father. I didnt believe it; but at 83½ (my present age) I find myself doing all sorts of silly things.

---

[123] Interview with Archibald Henderson, 'Bernard Shaw Talks of His "Saint Joan"', *Literary Digest International Book Review*, ii (Mar. 1924), 286. Brian Tyson, in *The Story of Shaw's 'Saint Joan'* (Kingston and Montreal: McGill-Queen's Univ. Press, 1982), gives a detailed account of Shaw's reliance on the trial records (that is, on T. Douglas Murray's book) in *Saint Joan*; and he also notes Shaw's consultation with a Roman Catholic priest of his acquaintance, Father Joseph Leonard. For correspondence about Joan's history between Shaw and Father Leonard in 1922, see *Collected Letters 1911–1925*, 795–800.

[124] Shaw to S. Cockerell, 27 Feb. 1924, *Collected Letters 1911–1925*, 867.

Thanks for the corrections.[125]

In the rehearsal copy of the play (also in the Harry Ransom Humanities Research Center) we can see emendations in Shaw's hand putting Calvin's death back a hundred years, and changing Charles's remark to Nell Gwynn, 'I do not remember my father,' to 'I am not a bit like him.'

Many of the historical details in the play suggest a familiarity with late seventeenth-century English history, and more particularly a familiarity with parts of Macaulay's *History of England*. Shaw uses well-known historical facts in order to invest Charles with impressive perspicacity. The play is set in 1680, eight years before the Revolution that forced James, who had become king upon Charles's death in 1685, to flee to France. Charles in the play knows that James will come to a bad end. He refers several times to William of Orange, and he warns James, 'They will have your head off inside of five years unless you jump into the nearest fishing smack and land in France' (Act I, *CPP* vii. 248). One of the more memorable passages in Macaulay's narrative is the description in his tenth chapter of James's flight to France, the culminating sentence of which is 'Soon after the dawn of Sunday the fugitives were on board of a smack which was running down the Thames.'[126] Later in Shaw's play Charles predicts that 'the Protestants will kill Jamie; and the Dutch lad will see his chance and take it. He will be king: a Protestant king.' It looks here as if Charles's prediction is not quite accurate, but in his next speech he comes closer to what actually happened in 1688: 'Jamie has just one chance. [The Protestants] may call in Orange Billy before they kill him; and then it will hardly be decent for Billy to kill his wife's father. But they will get rid of Jamie somehow . . .' (Act II, *CPP* vii. 291). Charles makes William seem less noble and generous than Macaulay does, but one does feel that Charles has had the advantage of reading an advance copy of *The History of England*.

This feeling is strengthened when Charles anticipates the Duke of Monmouth's rebellion in 1685 (Act II, *CPP* vii. 291), which is described in great detail in chapter v of Macaulay's *History*, and

---

[125] Shaw to G. M. Trevelyan, 20 Jan. 1940. MS in Harry Ransom Humanities Research Center, the University of Texas at Austin.
[126] Macaulay, *History of England*, ii. 339.

when he predicts that 'The Protestants will have you, Jamie, by hook or crook.... But they shall not have me. I shall die in my bed, and die King of England in spite of them' (Act I, *CPP* vii. 268); and that 'when I am dying . . . my last thought will be of Nelly' (Act I, *CPP* vii. 283). We have already seen that Shaw knew Macaulay's description of Charles's deathbed, and among Charles's last words in Macaulay's narrative is the injunction not to 'let poor Nelly starve'.[127] Another passage from Macaulay that Shaw evidently knew was the description in chapter iv of the flogging of Titus Oates in 1685. This time it is James who unconsciously alludes to the future in *Good King Charles*: when Charles brings up the subject of Oates, James's response is 'Flog him through the town. Flog him to death. They can if they lay on hard enough and long enough. The same mob that now takes him for a saint will crowd to see the spectacle and revel in his roarings' (Act I, *CPP* vii. 253).

One of Shaw's aims as a historical dramatist is to set the record straight. In the Preface to *Saint Joan* he demonstrates the errors and misconceptions of all previous literary treatments of the subject, and the play itself is offered as a corrective.[128] In the Preface to *Good King Charles* this play is offered as a corrective to the conventional undervaluation of Charles II: 'Unfortunately the vulgarity of his reputation as a Solomonic polygamist has not only obscured his political ability, but eclipsed the fact that he was the best of husbands. Catherine of Braganza, his wife, has been made to appear a nobody, and Castlemaine, his concubine, almost a great historical figure. When you have seen my play you will not make that mistake, and may therefore congratulate yourself on assisting at an act of historical justice' (*CPP* vii. 206–7). In *Good King Charles*, then, Shaw is attempting historical accuracy.

This accuracy, however, is not a matter of mere facts. 'Now the facts of Charles's reign', Shaw observed in the Preface, 'have been

---

[127] Macaulay, *History of England*, i. 342 (chap. iv).

[128] 'J'ai écrit cette pièce comme un acte de justice et de piété envers Jeanne outrageusement traitée par Shakespeare, S[c]hiller, Voltaire, Anatole France, de même que Barbier et autres dramaturges de seconde zone' (Shaw to René Viviani, 13 Mar. 1924, *Collected Letters 1911–1925*, 870. Cf. his comment a month later that 'Mahomet, like Joan, needs to be rescued from Voltaire, whose play about him is really an outrage. I thought of him as a play subject at one time' (Shaw to Hesketh Pearson, 13 Apr. 1924, ibid. 875).

chronicled so often by modern historians of all parties, from the Whig Macaulay to the Jacobite Hilaire Belloc, that there is no novelty left for the chronicler to put on the stage.' The trick is to look beyond the facts: 'But when we turn from the sordid facts of Charles's reign, and from his Solomonic polygamy, to what might have happened to him but did not, the situation becomes interesting and fresh' (*CPP* vii. 203–4). This type of history play suggests another side of Shaw's treatment of history on the stage. In his programme note for *Saint Joan* Shaw defends most of the play as factually accurate, and adds that 'The Epilogue is obviously not a representation of an actual scene, or even of a recorded dream; but it is none the less historical' (*CPP* vi. 213). Clearly, the term 'historical' means more to Shaw than 'factual'. Shaw regards ideas rather than events as the essence of history, and his dramatic practice reflects this attitude.

When I quoted from Shaw's 1899 programme note for *Caesar and Cleopatra* a few pages earlier, I deliberately omitted its final sentence. After his references to Mommsen, Warde Fowler, St George Stock, and Murray's *Handbook for Egypt*, Shaw goes on to say: 'Many of these authorities have consulted their imaginations, more or less. The author has done the same.' We have in this programme note a typical example of the way in which Shaw likes to occupy both sides of the field. On the one side, he gives the audience an imposing list of authorities. On the other side, he undercuts the list and reveals the whole note as a spoof—but still leaving the impression that he is aware of all sorts of respectable authorities, even if he is above relying on them.

To have suggested at the end of the nineteenth century that 'Many of these authorities have consulted their imaginations, more or less,' and 'The author has done the same' is to dissociate oneself radically from the prevailing trends in English academic history at the time, and in *Everybody's Political What's What?* Shaw explicitly rejects the assumptions of the Rankean school of history. 'When the German so-called Historical School in the nineteenth century repudiated classical, dramatized, apriorist history, and called for masses of recorded facts and years of dryasdust searches through libraries for documents, they were overlooking the cardinal fact that their method is physically impossible, because most of the facts are hidden or out of reach, and such records as exist are mostly lies, or at best wishful guesses.'[129]

Implicitly, Shaw rejected the Rankean position throughout his career. In 1894, during the first run of *Arms and the Man* in London, he provided an important statement of his dramatic historiography in a newspaper interview. He was asked whether he considered that a historical play is bound to be 'substantially accurate as to facts, etc.', and he replied:

> Not more so than any other sort of play. Historical facts are not a bit more sacred than any other class of facts. In making a play out of them you must adapt them to the stage, and that alters them at once, more or less. Why, you cannot even write a history without adapting the facts to the conditions of literary narrative, which are in some respects much more distorting than the dramatic conditions of representation on the stage. Things do not happen in the form of stories or dramas; and since they must be told in some such form, all reports, even by eyewitnesses, all histories, all stories, all dramatic representations, are only attempts to arrange the facts in a thinkable, intelligible, interesting form—that is, when they are not more or less intentional efforts to hide the truth, as they very often are.

He then explained how in writing *Arms and the Man* he completed the substance of the play and added the setting later—the Serbo-Bulgarian War of 1885–6. He consulted his friend Sergius Stepniak about the Bulgarian background, and 'followed the facts he gave me as closely as I could, because invented facts are the same stale stuff in all plays, one man's imagination being much the same as another's in such matters, whilst real facts are fresh and varied'. He concluded his answer to the question by saying, 'So you can judge exactly how far my historical conscience goes. If I were to write a play about Julius Caesar, it would not really be historical; but I should take care not [to] let him appear with a revolver and a field-glass all the same' ('Ten Minutes with Mr Bernard Shaw', *CPP* i. 480–2).

The historical facts can be desirable in a history play, but they are not essential. Just as Shaw claimed to have filled in the local details at a late stage in the composition of *Arms and the Man*, so he informed Ellen Terry while he was working on *The Devil's Disciple* that 'The play only exists as a tiny scrawl in my note

---

[129] *Everybody's Political What's What?*, 365.

books—things I carry about in my pockets. I shall have to revise it & work out all the stage business, besides reading up the history of the American War of Independence before I can send it to the typist to be readably copied.'[130] In the case of *The Man of Destiny*, Shaw told the actor Richard Mansfield that he had modelled Napoleon on him. 'I studied the character from you, and then read up Napoleon and found that I had got him exactly right.'[131] There is no way of knowing just how accurate such comments are, but we do know that three days after completing *The Man of Destiny* Shaw asked the publisher T. Fisher Unwin to send him a life of Napoleon that was being run serially in an American magazine.[132]

There are a number of other places in which Shaw gives the impression that he is scornful of factual accuracy in his history plays. In the Notes to *Caesar and Cleopatra*, he introduced some information that he had from Gilbert Murray about Greek medications by observing that Murray, 'as a Professor of Greek, has applied to classical antiquity the methods of high scholarship (my own method is pure divination)' (*CPP* ii. 294). At the time he began the play, he denied that he had 'been reading up Mommsen —and people like that'. History, he argued, 'is only a dramatisation of events. . . . I never worry myself about historical details until the play is done; human nature is very much the same always and everywhere. And when I go over my play to put the details right I find there is surprisingly little to alter. . . . You see, I know human nature. Given Caesar, and a certain set of circumstances, I know what would happen, and when I have finished the play you will find I have written history.'[133] Of *Androcles and the Lion* he wrote, in a programme note for the New York production in 1915: 'The play is probably as true to history as it is in the nature of a good play to be' ('A Note to "Androcles and the Lion"', *CPP* iv. 650). In his programme note for the 1913 London production he took up this subject at greater length. He acknowledged that

---

[130] Shaw to Ellen Terry, 30 Nov. 1896, *Collected Letters 1874–1897*, 706.
[131] Shaw to Richard Mansfield, 8 Sept. 1897, *Collected Letters 1874–1897*, 803.
[132] Shaw to T. Fisher Unwin, 27 Aug. 1895, *Collected Letters 1874–1897*, 551–2. This biography was published in book form in four volumes in 1901, as *Life of Napoleon Bonaparte*; the author was William Milligan Sloane, a historian at Princeton University.
[133] [Clarence Rook], 'Mr. Shaw's Future. A Conversation', *The Academy* (30 Apr. 1898), 476.

Introduction 51

'In representing a Roman centurion and a Roman captain (a pure invention) as corresponding to a British sergeant and a British company officer, some violence may or may not have been done to the petty accuracies of military history,' and he said he did not much care whether he used the Roman military titles that were precisely appropriate to his characters. His conclusion is one of his most amusing and important statements on the subject we are looking at here.

> In short, if you demand my authorities for this and that, I must reply that only those who have never hunted up the authorities as I have believe that there is any authority who is not contradicted flatly by some other authority. Marshal Junot, reproached for having no respect for ancestry, said that he was an ancestor himself. In the same spirit I point out that the authorities on the story of Androcles and on the history of the early Christian martyrs are the people who have written about them; and now that I, too, have written about them, I take my place as the latest authority on the subject and ask you to respect me accordingly.
> (*CPP* iv. 583–4.)

In the first part of this section we saw that in Shaw's view economic considerations have some importance in history, but ideas are much more important. Now we see that facts have some importance in historical writing, but it is ideas that really count. It would be far less misleading if the standard historical works of reference were wrong as to the facts of Joan's history, 'and right in their view of the facts', Shaw said in the Preface to *Saint Joan* (*CPP* vi. 66). A historian's view of the facts involves his imagination, and in the writing of history, as in the historical process itself, 'imagination is the beginning of creation'. In 1893, in one of his weekly articles on music, Shaw wrote about Wagner's critical work *Opera and Drama*, which was appearing in English translation at that time. His comments are relevant to his conception of historical writing, and to his conception of his own history plays (which were still to be written). Wagner's *Opera and Drama*, he said, 'did more than any other writing of Wagner's to change people's minds on the subject of opera'. Then he went on to compare it with other great works of the nineteenth century:

> Like all the books which have this mind-changing property—Buckle's History of Civilization, Marx's Capital, and Ruskin's Modern Painters

are the first instances that occur to me—it professes to be an extraordinarily erudite criticism of contemporary institutions, and is really a work of pure imagination, in which a great mass of facts is so arranged as to reflect vividly the historical and philosophical generalizations of the author, the said generalizations being nothing more than an eminently thinkable arrangement of his own way of looking at things, having no objective validity at all, and owing its subjective validity and apparent persuasiveness to the fact that the rest of the world is coming round by mere natural growth to the author's feeling, and therefore wants 'proof,' historical, philosophical, moral, and so on, that it is 'right' in its new view.[134]

Historical truth, like any other kind of truth, is not essentially a question of fact but of interpretation, and to interpret is to imagine.

Shaw's use of facts in his history plays varies a good deal, because the plays themselves have markedly different relationships to the historical record. At one end of the spectrum there are *Saint Joan*, subtitled 'A Chronicle Play', and *Caesar and Cleopatra*, which is subtitled 'A History'—a term that Shaw uses to mean 'chronicle'. Each of these plays presents historical characters in authentic historical situations that had previously been well recorded. 'The "histories" of Shakespear are chronicles dramatized; and my own chief historical plays, Caesar and Cleopatra and St Joan, are fully documented chronicle plays of this type. Familiarity with them would get a student safely through examination papers on their periods' (Preface to *Good King Charles*, *CPP* vii. 203). When *Caesar and Cleopatra* was revived in London in 1913, Shaw wrote that the technically interesting part of his play was 'that it is "a history": the old term for a chronicle play'. There have been many plays since Shakespeare's time in which historical characters figure in invented dramatic incidents, but there have been almost no examples of 'the real thing, the play in which the playwright simply takes what the chronicler brings him and puts it on the stage just as it is said to have happened' ('Caesar and Cleopatra, by the Author of the Play', *CPP* ii. 311). A few months after he had completed the composition of the play, he told William Archer that he had found the chronicle form constricting, and he

---

[134] *Shaw's Music*, iii. 18–19.

blamed 'the scatteriness' of the play on the exigencies of the form. 'The defects of "C & C" seem to me to be inherent in the *genre* Chronicle Play.... The chronicle ties you to the exposition of Caesar's position at Alexandria;[135] and there is no drama in it because Caesar was so completely superior to his adversaries that there was virtually no *conflict*, only a few *adventures*, chiefly the hairbreadth escape when he jumped into the harbor. I was desperate about the business until, like Columbus with the egg, I solved the problem by making Cleopatra commit a murder.' Shaw also believed that he had succeeded in making Caesar a great man, as opposed to a nobody who is arbitrarily labelled 'Caesar'. 'That achieved, I give up the rest as hopeless.' The problem is that the chronicle 'must tell its historical story'. Shaw discovered in writing the play that

the peculiar characteristics of the Shakespear chronicle play are not due to his neglect or failure to construct them like Othello, but are produced by the technical conditions of the feat. You say in the chronicle play, 'I will accept character and story from outside the drama—from History, not from my own dramatic invention & the needs of the dramatic appetite; and I will make the best play I can out of them. In the Othello–Devil's Disciple genre, you make the whole thing—character, story & everything else—out of the tree in your own garden.[136]

The other history play of Shaw's that is closest to the two chronicle plays is a late piece, *The Six of Calais*, a one-act play depicting Edward III, Queen Philippa, and others in a historical incident in 1347. Its subtitle is 'A Medieval War Story', but in the first proof the subtitle was fuller and more revealing: 'A Medieval War Story Told by Froissart and Now Retold with Certain Necessary Improvements by a Fellow of the Royal Society of Literature'.[137] In the published version, Shaw works this idea into his Preface: 'I have had to improve considerably on the story as told by that absurd old snob Froissart.... He made a very poor job of it in my opinion' (*CPP* vi. 975). *The Man of Destiny* takes

---

[135] Cf. Shaw's comment to Archer in 1916 that *Caesar and Cleopatra*, as a chronicle play, 'is nailed down to history' (30 Dec. 1916, *Collected Letters 1911–1925*, 445).
[136] Shaw to W. Archer, 27 July 1899, *Collected Letters 1898–1910*, 93–4.
[137] Proof copy of *The Six of Calais* in Harry Ransom Humanities Research Center, the University of Texas at Austin; Dan H. Laurence, *Bernard Shaw: A Bibliography* (2 vols., Oxford: Clarendon, 1983), i. 400.

us a little further away from the historical record. The main character is historical, and so is the setting—an inn on the road from Lodi to Milan, two days after the Battle of Lodi—but the incident that forms the plot of the play is pure invention. The apt subtitle of this play, 'A Fictitious Paragraph of History', would apply equally to *The Dark Lady of the Sonnets* and *Great Catherine*, which combine historical characters and fictitious plots. Shaw certainly did not make any large claims for *Great Catherine*: 'I must not . . . pretend', he wrote in his Preface, 'that historical portraiture was the motive of a play that will leave the reader as ignorant of Russian history as he may be now before he has turned the page' (*CPP* iv. 899).

Then we move still further from factual history in *The Devil's Disciple*, the subtitle of which—'A Melodrama'—indicates that it is only superficially a history play. Here we have a historical setting, New England during the American Revolution, and two historical characters, Burgoyne and (up to a point) Brudenell. Then, as Shaw says in his Notes to the play, 'The rest of the Devil's Disciple may have actually occurred, like most stories invented by dramatists; but I cannot produce any documents' (*CPP* ii. 150). In fact, the rest of the play is a melodrama that happens to have a historical setting. The play is less of a history play than several of Shaw's plays that would not usually be given this designation (such as *Major Barbara* and *John Bull's Other Island*). I should also mention here *The Glimpse of Reality*, a trifle written in 1909–10 with a Renaissance Italian setting, but nothing else that is historical about it.

The subtitle of '*In Good King Charles's Golden Days*' is 'A True History that Never Happened'. One of Shaw's best and most interesting history plays, it offers a historical setting, historical characters, and a dialogue that rises above historical incidents into the realm of historical ideas. For this play Shaw did make large historical claims, and quite rightly. A month after he began composing the play he told its first producer that what he was 'aiming at so far is an educational history film. The people will wear XVII century costumes regardless of expense, numbers, and salaries.'[138] But the historical veracity of the play does not lie in its

---

[138] Shaw to R. Limbert, 24 Dec. 1938. MS in Harry Ransom Humanities Research Center, the University of Texas at Austin.

detail. In *Everybody's Political What's What?* Shaw provides a good indication of the true nature of the play:

> At last I became a historian myself. I wrote a play entitled In Good King Charles's Golden Days. For the actual occurrence of the incidents in it I cannot produce a scrap of evidence, being quite convinced that they never occurred; yet anyone reading this play or witnessing a performance of it will not only be pleasantly amused, but will come out with a knowledge of the dynamics of Charles's reign: that is, of the political and personal forces at work in it, that ten years of digging up mere facts in the British Museum or the Record Office could not give.[139]

As Martin Meisel has noted, *Good King Charles* carries Shaw's earlier principles and practice to a logical conclusion, in that 'external veracity is altogether abandoned for the sake of essential truth'.[140] For a play to be truly historical, in Shaw's view, it must convey the essential forces of an age, the currents of powerful ideas, the dynamics of an epoch. Even in his 'chronicle plays' the historical record is subordinated to the wider concerns of the play. *Saint Joan* dramatizes the dynamics of the Middle Ages and the Renaissance, and *Caesar and Cleopatra* is not essentially a record of the facts of Caesar's Alexandrian campaign, but rather a dramatic exhibition of a historical hero.

[139] *Everybody's Political What's What?*, 181.
[140] Meisel, *Shaw and the Nineteenth-Century Theater*, 376–7.

2

The Heroic in History

Charles II, in *In Good King Charles's Golden Days*, is in the line of Shaw's great men in history, and he is also in a longer line, one that extends back at least to the late eighteenth century. 'The worship of God is: Honouring his gifts in other men, each according to his genius, and loving the greatest men best: those who envy or calumniate great men hate God; for there is no other God',[1] wrote Blake in the early 1790s, and there is a tradition of hero-worship running through the literature of the nineteenth century. Walter Houghton, who devotes a chapter of *The Victorian Frame of Mind* to this subject, says that 'In no other age were men so often told to take "the great ones of the earth" as models for imitation, or provided with so many books with titles like *Heroes and Hero-Worship, Lectures on Great Men, A Book of Golden Deeds, The Red Book of Heroes*.'[2] In a period when for many people the worship of God was becoming untenable, the worship of man became a substitute. Heroes were found in legend, in history, and in contemporary life; and King Arthur, Sir Walter Raleigh, and the Duke of Wellington became fit objects of veneration, inspiration, and imitation.

The Victorian writer who most obviously represents this attitude is Carlyle. In May 1840 he gave a series of six lectures that he published in the following year as *On Heroes, Hero-Worship, and the Heroic in History*. Taking Odin, Mahomet, Dante, Shakespeare, Luther, John Knox, Samuel Johnson, Rousseau, Burns, Cromwell, and Napoleon as his heroes, he declared that 'Universal History, the history of what man has accomplished in this world, is at bottom the History of the Great Men who have worked here'. One of the functions of the Great Man, according to Carlyle, is to inspire others to higher thoughts and achieve-

---

[1] William Blake, *The Marriage of Heaven and Hell, The Complete Writings of William Blake*, ed. Geoffrey Keynes (London: Oxford Univ. Press, 1966), 158 (Plates 22–4).
[2] Walter E. Houghton, *The Victorian Frame of Mind* (New Haven and London: Yale Univ. Press, 1957), 305. The chapter on 'Hero Worship' is on 305–40.

The Heroic in History

ments: 'We cannot look, however imperfectly, upon a great man, without gaining something by him.'[3] The worship of the Great Man is an essentially religious act. 'Reverence for Human Worth, earnest devout search for it and encouragement of it, loyal furtherance and obedience to it: this, I say, is the outcome and essence of all true "religions," and was and ever will be.'[4] In his *Heroes and Hero-Worship*, as in his portrait of Mirabeau in *The French Revolution* and in his large-scale accounts of Cromwell and Frederick the Great, Carlyle wanted to restore the religious impulse in a godless generation. He wanted to reveal the divine principle working in history through great individuals, so that (as he said in introducing his *Frederick the Great*) 'the modern Nations may again become a little less godless, and again have their "epics"'.[5]

Shaw considered writing plays about two of Carlyle's heroes, Cromwell and Mahomet, and the first history play he actually did write was about another of them, Napoleon. (I omit *The Dark Lady of the Sonnets*, in that its Shakespeare is not in any sense a Carlylean hero.) Furthermore, the encouragement of 'reverence for Human Worth' is one of Shaw's intentions in such history plays as *Caesar and Cleopatra* and *Saint Joan*, and even *Good King Charles*. One of the defects of Shakespeare and Dickens as artists, Shaw argued in the Epistle Dedicatory to *Man and Superman*, is their inability to balance their exposures of human weakness and folly 'with any portrait of a prophet or a worthy leader' (*CPP* ii. 520). The value of such a portrait, for Shaw as for Carlyle, is religious, in that it gives people a reason for living and may inspire them to reach beyond their present level of attainment. 'The apparent freaks of nature called Great Men mark not human attainment but human possibility and hope. They prove that though we in the mass are only child Yahoos it is possible for creatures built exactly like us, bred from our unions and developed from our seeds, to reach the heights of these towering heads' (Preface to *Geneva*, *CPP* vii. 41).

---

[3] Thomas Carlyle, *On Heroes, Hero-Worship, and the Heroic in History*, The *Works of Thomas Carlyle*, ed. H. D. Traill (Centenary Edn., 30 vols., London: Chapman and Hall, 1896–9), v. 1 ('The Hero as Divinity').
[4] Carlyle, 'Downing Street', *Latter-Day Pamphlets*, *Works*, xx. 106.
[5] Carlyle, *History of Friedrich II of Prussia Called Frederick the Great*, i. 19, *Works* xii (Book i, chap. 1).

While in his emphasis on the Great Man, Shaw appears to be closer to Carlyle than to Macaulay,[6] there is one respect in which his heroes are more akin to Macaulay's. The virtues and vices of Shaw's characters do not come in sets, and it is one of Shaw's observed principles of human character that genius in one area does not imply any special quality in other areas. (Louis Dubedat, the gifted artist and moral scoundrel in *The Doctor's Dilemma*, is one notable example in the plays.) To present mixed characters as one's heroes is not only true to experience, but in Shaw's view it is dramatically necessary on the modern stage. Contemporary audiences demand heroes who, 'instead of walking, talking, eating, drinking, sleeping, making love and fighting single combats in a monotonous ecstasy of continuous heroism, are heroic in the true human fashion: that is, touching the summits only at rare moments, and finding the proper level of all occasions, condescending with humour and good sense to the prosaic ones as well as rising to the noble ones, instead of ridiculously persisting in rising to them all on the principle that a hero must always soar, in season and out of season' ('Bernard Shaw and the Heroic Actor', *CPP* ii. 307). Carlyle's heroes do not exactly walk, talk, eat, etc. in a monotonous ecstasy of continuous heroism, but they are much closer to this state than Shaw's are. Had Carlyle attempted to write a play about Cromwell, one suspects it would have shown him in a purely heroic light. Had Shaw written his proposed 'Death of Cromwell', we would almost certainly have seen the man's weaknesses and failures as well as his strengths and successes. That is the way Shaw treats all of his characters, including the heroes of history.

The hero Shaw has specifically in mind in the quotation in the previous paragraph is his Julius Caesar. A year after writing *Caesar and Cleopatra*, he described it as 'the first & only adequate dramatization of the greatest man that ever lived. I want to revive, in a modern way and with modern refinement, the

---

[6] In his letter to Winston Churchill, one of Shaw's objections to Macaulay is that he 'did not reckon with genius. I won't say that he didn't understand it: nobody understands it; but some writers recognise it. Carlyle did, for instance: he was so good at heroes that when his stock ran short he began manufacturing heroes out of mere *arrivistes*. But Macaulay could not tackle even such obvious cases as Byron and Bunyan' (Shaw to Winston Churchill, 8 May 1934, in Martin Gilbert, *Winston S. Churchill*, v, Companion Part 2, *The Wilderness Years 1929–1935* (London: Heinemann, 1981), 785).

sort of thing that Booth did the last of in America: the projection on the stage of the hero in the big sense of the word.'[7] Nevertheless, the play on the whole avoids excessive solemnity; the hero is knocked off his pedestal to reveal the mortal:

CLEOPATRA. I am going to dress you, Caesar. Sit down. [*He obeys*] These Roman helmets are so becoming! [*She takes off his wreath*] Oh! [*She bursts out laughing at him*].
CAESAR. What are you laughing at?
CLEOPATRA. Youre bald [*beginning with a big B, and ending with a splutter*].
CAESAR [*almost annoyed*]. Cleopatra! ...

(Act II, *CPP* ii. 221.)

This scene is deliberately reminiscent of Cleopatra's arming of Antony in Act IV, scene iv of Shakespeare's *Antony and Cleopatra*, and it also has behind it the whole epic tradition of the arming of the hero. The scene therefore frustrates the audience's expectation of a piece of conventional heroic display. We are reminded of Caesar's mortal nature in a number of anti-climactic moments of this type, and also in the fact that he is leaving his conquests in Alexandria to return to Rome where he will be assassinated. Caesar is not only the magnificent hero; he is an ageing man at the end of his career.

This is not to deny that Caesar *is* the magnificent hero. When he is knocked off the pedestal, it is the greatness as well as the limitation of the mortal that is revealed. In his wisdom, and in his military ability, he towers over everyone else in the play, Egyptian and Roman alike. Shaw told Hesketh Pearson in 1918 that the Sphinx scene 'was suggested by a French picture of the Flight into Egypt', and critics have proposed that Shaw's use of a painting by Luc Olivier Merson of Christ and the Virgin Mary in the arms of the Sphinx indicates Caesar as a precursor of Christ.[8] The other

---

[7] Shaw to Mrs Richard Mansfield, 3 May 1899, *Collected Letters 1898–1910*, ed. Dan H. Laurence (London: Max Reinhardt, 1972), 90.

[8] Hesketh Pearson, *G.B.S.: A Full Length Portrait* (New York and London: Harper and Brothers, 1942), 187; Stanley Weintraub, *The Unexpected Shaw* (New York: Frederick Ungar, 1982), 70. James Anthony Froude, the Victorian historian and biographer of Carlyle, wrote a book about Caesar in which he suggested that 'it may be said that he came into the world at a special time and for a special object'—that is, to prepare the way for the birth of Christianity (*Caesar: A Sketch* (London: Longmans, 1886), 558–9).

evidence for this connection would be Caesar's principle of clemency, and also the religious language of Ra's Prologue. One should be cautious, however, in considering Caesar as a religious or prophetic figure. Ra's Prologue, which asserts that 'Julius Caesar was on the side of the gods' and represents him as participating in a divine will larger than his own (*CPP* ii. 163–4), was written in 1912, fourteen years after the play itself, and it expresses a later stage in Shaw's thinking.[9] Within the 1898 *Caesar and Cleopatra*, Caesar's role is largely that of the adroit military commander, who is able to maintain the power of the Roman empire in Egypt against great odds. Like Napoleon in *The Man of Destiny*, Caesar is the imperialist, and he is upholding the status quo in the world. Caesar and Joan may seem comparable as Shavian heroes, but (to leave aside all the other differences) Joan is resisting an 'imperialist' occupation force, and challenging the status quo.

An illuminating way to look at *Caesar and Cleopatra* is to see it as an irreverent, subversive response to Shakespeare's *Antony and Cleopatra* and *Julius Caesar*.[10] 'As to my alleged failure to present the erotic Caesar,' Shaw wrote in response to adverse criticisms of his play in 1913, 'that is a matter almost too delicate for discussion. But it seems to me that the very first consideration that must occur to any English dramatic expert in this connection is that Caesar was not Antony' ('Caesar and Cleopatra, by the Author of the Play', *CPP* ii. 314). Caesar is like Joan in this one respect at least: both are beyond sexual involvement. And the very first consideration that will occur to the dramatic expert with respect to Cleopatra is that she is not Shakespeare's Cleopatra. Instead of a mature, tragic woman we are presented with a silly girl. It is true that she matures slightly in the course of the play, under Caesar's tutelage, but at the end of the last act she makes the choice that most decisively reveals her inferior nature.

CAESAR. . . . Come, Cleopatra: forgive me and bid me farewell; and I will send you a man, Roman from head to heel and Roman of the noblest; not old and ripe for the knife; not lean in the arms and cold in

[9] See J. L. Wisenthal, 'Shaw and Ra: Religion and Some History Plays', *SHAW: The Annual of Bernard Shaw Studies*, 1 (1981), 45–56.
[10] See my Introduction to the *Man of Destiny* and *Caesar and Cleopatra* volume of Shaw, *Early Texts: Play Manuscripts in Facsimile*, ed. Dan H. Laurence (New York and London: Garland, 1981), pp. xiv–xv.

the heart; not hiding a bald head under his conqueror's laurels; not stooped with the weight of the world on his shoulders; but brisk and fresh, strong and young, hoping in the morning, fighting in the day, and revelling in the evening. Will you take such an one in exchange for Caesar?
CLEOPATRA [*palpitating*]. His name, his name?
CAESAR. Shall it be Mark Antony? [*She throws herself into his arms*].
RUFIO. You are a bad hand at a bargain, mistress, if you will swop Caesar for Antony.

(*CPP* ii. 291–2.)

The full measure of Cleopatra's inferiority is her preference to be in Shakespeare's play instead of Shaw's. To the true hero in this play sexual involvement, like other personal pleasure, is negligible or altogether absent. Cleopatra has nothing of the heroic in her.

The principal reason that Shaw found it necessary to provide the first and only adequate dramatization of the greatest man who ever lived is that Shakespeare had libelled him. '[H]ero-worshippers have never forgiven [Shakespeare] for belittling Caesar and failing to see that side of his assassination which made Goethe denounce it as the most senseless of crimes', Shaw wrote in the Preface to *The Dark Lady of the Sonnets* (*CPP* iv. 298). Three months before he started writing *Caesar and Cleopatra*, he began his review of a production of *Julius Caesar* in the *Saturday Review* with the cry, 'The truce with Shakespear is over', and he had some particularly hard words for the play's treatment of Caesar: 'It is impossible for even the most judicially minded critic to look without a revulsion of indignant contempt at this travestying of a great man as a silly braggart, whilst the pitiful gang of mischief-makers who destroyed him are lauded as statesmen and patriots. There is not a single sentence uttered by Shakespear's Julius Caesar that is, I will not say worthy of him, but even worthy of an average Tammany boss.'[11] In an article in 1907, Shaw argued that Shakespeare, in following Plutarch's republican inclinations, was a man of his age ('Bernard Shaw and the Heroic Actor', *CPP* ii. 309–10); and as we shall see in the next chapter, Shaw used Shakespeare's sympathy with Brutus as a

---

[11] Shaw, *Our Theatres in the Nineties* (3 vols., London: Constable, 1954), iii. 297, 298.

touchstone for English political attitudes over the last three hundred years.

Julius Caesar's portrait of its title character is also defective because it is based on Shakespeare's 'essentially knightly conception of a great statesman-commander', a conception that was challenged in the nineteenth century by Mommsen and Carlyle. Mommsen 'explains the immense difference in scope between the perfect knight Vercingetorix and his great conqueror Julius Caesar', while Carlyle, 'with his vein of peasant inspiration, apprehended the sort of greatness that places the true hero of history so far beyond the mere *preux chevalier*, whose fanatical personal honor, gallantry, and self-sacrifice, are founded on a passion for death born of inability to bear the weight of a life that will not grant ideal conditions to the liver. This one ray of perception became Carlyle's whole stock-in-trade; and it sufficed to make a literary master of him' (*CPP* ii. 45–6). Shaw follows Carlyle in dramatizing a hero who does not adhere to any code but is original in his morality, and whose leading characteristics are vitality of mind and intensity of will.

These same heroic qualities are to be found in Shaw's other historical-hero play of the 1890s: *The Man of Destiny*, and in this play too Shaw sets out to put a real hero on the stage in the place of another dramatist's defective job. Here the other dramatist is Victorien Sardou, whose Napoleon play, *Madame Sans-Gêne*, he saw in 1895 while he was working on *The Man of Destiny*. Two years later he had an opportunity to denounce the Sardou play publicly, when Sir Henry Irving produced it at the Lyceum Theatre. 'Sardou's Napoleon', he commented, 'is rather better than Madame Tussaud's, and that is all that can be said for it. . . . He is nothing but the jealous husband of a thousand fashionable dramas, talking Buonapartiana.' Since Sardou's Napoleon is just a conventional stage figure with superficial Napoleonic touches, all one would need to do is change the costume and local references and one would have a stage Julius Caesar.[12] Sardou left out everything that was truly Napoleonic and presented only the trappings. Shaw in his play wished to leave out the trappings and to present something of Napoleon's spirit and will, the human qualities that raise a Man of Destiny above the ordinary jealous

[12] Shaw, *Our Theatres in the Nineties*, iii. 110.

husband. In his refusal to take any interest in his wife's alleged infidelity, Shaw's Napoleon rises above the mechanisms of the well-made play, and he displays his greatness on the stage without the grandeur of imperial France on which *Madame Sans-Gêne* relies for its setting. In Shaw's play we have, instead of *l'Empereur* at the height of his power, the young officer of the Italian campaign, in the early stages of his rise to greatness. (Similarly, Shaw's Julius Caesar is shown at a somewhat earlier stage of his career than the one that Shakespeare's play has made familiar, and the title of the proposed Cromwell play suggests a comparable perspective—Cromwell *after* the height of his power.)[13]

In *The Man of Destiny*, Napoleon's opponent is not of the military sort, but his role as military conqueror is none the less prominent in the play. In his victory over the Strange Lady, he displays the same qualities of will, daring, and ruthlessness that won him the Battle of Lodi two days earlier. The play invites its audience to admire Napoleon the military conqueror. In the initial 'stage direction', which is really a brief essay on the central character, readers of the play are provocatively informed that '*Napoleon, as a merciless cannonader of political rubbish, is making himself useful: indeed, it is even now impossible to live in England without sometimes feeling how much that country lost in not being conquered by him as well as by Julius Caesar*' (*CPP* i. 608). In other places Shaw suggested that the defeat of Napoleon by Wellington was unfortunate; in 1898 in *The Perfect Wagnerite*, for example, he wrote: 'It seems hardly possible that the British army at the battle of Waterloo did not include at least one Englishman intelligent enough to hope, for the sake of his country and humanity, that Napoleon might defeat the allied sovereigns.'[14]

This admiration for Napoleon is on the whole an early view, however, and after the First World War Shaw's treatment of him was markedly different. The Preface to *Heartbreak House*, written a few months after the end of the war, has a reference to 'Napoleon and all the other scourges of mankind' (*CPP* v. 31), and the Preface to *The Millionairess*, written in 1935, describes

---

[13] In this paragraph I have drawn on my Introduction to the Garland facsimile edition of *The Man of Destiny* and *Caesar and Cleopatra*.
[14] *Shaw's Music*, ed. Dan H. Laurence (3 vols., London: Max Reinhardt, The Bodley Head, 1981), iii. 466–7.

him as 'a very ordinary snob in his eighteenth-century social outlook'—a far cry from the independent-minded man of destiny of 1895. This Preface is also disparaging about Napoleon's military abilities, describing him as a commander who could only make 'the textbook moves he had learnt at the military academy'. 'Unfortunately for himself and Europe Napoleon was fundamentally a commonplace human fool' is Shaw's judgement in 1935, and the Preface to *The Millionairess* concerns itself with the wider problems raised by the success of such a man: 'But the vulgarer fool and the paltrier snob you prove Napoleon to have been, the more alarming becomes the fact that this shabby-genteel Corsican subaltern (and a very unsatisfactory subaltern at that) dominated Europe for years, and placed on his own head the crown of Charlemagne. Is there really nothing to be done with such men but submit to them until, having risen by their specialities, they ruin themselves by their vulgarities?' (*CPP* vi. 855–7). *Everybody's Political What's What?*, published in 1944, makes the point that Napoleon is unusual among military leaders in that he glorified war, and it has a disparaging discussion of his behaviour on St Helena after his defeat.[15]

Tragedy of an Elderly Gentleman, the fourth part of *Back to Methuselah*, was begun in the last year of the War and completed two years later.[16] One of its characters is Cain Adamson Charles Napoleon, Emperor of Turania—the type of the military hero. As his name suggests, he is an amalgam of the original murderer Cain, Charles XII of Sweden (1682–1718), and Napoleon. His first speech in the play, 'I am the Man of Destiny', implies that he is principally the Napoleonic man, and his name in the text is Napoleon. Like Shaw's 1895 Napoleon, he must match his will against that of a strange lady—in this case a veiled woman who is to act as the Oracle. The 1918–20 Napoleon, however, is quite unlike his 1895 counterpart. He is only a short-liver, and therefore feeble in comparison with the Oracle; thus all of his Napoleonic talk appears pretentious and absurd.

---

[15] Shaw, *Everybody's Political What's What?* (London: Constable, 1944), 129, 338–9.
[16] It was written between 21 May 1918 and 15 Mar. 1920 (British Library Additional MS 50631, fos. 94, 115).

NAPOLEON [*folding his arms*]. I am not intimidated: no woman alive, old or young, can put me out of countenance. Unveil, madam. Disrobe. You will move this temple as easily as shake me.
THE ORACLE. Very well [*she throws back her veil*].
NAPOLEON [*shrieking, staggering, and covering his eyes*]. No. Stop. Hide your face again. [*Shutting his eyes and distractedly clutching at his throat and heart*] Let me go. Help! I am dying.
THE ORACLE. Do you still wish to consult an older person?
NAPOLEON. No, no. The veil, the veil, I beg you.
THE ORACLE [*replacing the veil*]. So.
NAPOLEON. Ouf! One cannot always be at one's best.

(Act II, *CPP* v. 533–4.)

After this display, it is difficult to be properly impressed by the Man of Destiny's Carlylean utterances about the value of his life as a Great Man: '[T]he value of human life is the value of the greatest living man. . . . I matter supremely. . . . If you kill me, or put a stop to my activity (it is the same thing), the nobler part of human life perishes' (Act II, *CPP* v. 538). His account of his work as a military hero gives a very different impression from the one we gain in *The Man of Destiny*. His sole talent, he explains to the Oracle, is to organize slaughter, 'to give mankind this terrible joy which they call glory; to let loose the devil in them that peace has bound in chains' (Act II, *CPP* v. 535). He realizes that war is appalling, but knows that he can continue to be a Great Man only by exercising his fatal talent. He ends up 'gibbering impotently' at the base of the monument erected to Falstaff to honour his praise of cowardice.[17]

This transformed view of Napoleon reflects a shift in Shaw's attitude towards the Great Man. Even at the beginning of his career, in *Arms and the Man* (1893–4), the conventional operatic military hero is exposed as a fraud, and the prosaic, bourgeois Bluntschli is shown to be the superior man. We could consider this conception of the Hero as Hotel-Keeper as an implicit challenge to Carlyle's view of things; Bluntschli is not even a heroic Captain of Industry. Andrew Undershaft in *Major Barbara*

---

[17] For a discussion of Napoleon that is similar to the dramatization in *Back to Methuselah*, see Shaw's letter to H. G. Wells, 1 July 1921, in response to Wells's account in *The Outline of History* (*Collected Letters 1911–1925*, ed. Dan H. Laurence (London: Max Reinhardt, 1985), 724–5). Instead of sending this letter, Shaw sent Wells the Napoleon scene from his play.

(1905) is closer to the Carlylean Captain of Industry, and his trade is connected with war, but he is not the military hero. His concerns are mainly economic; and as for war, he does not care which side wins. In *Major Barbara*, war largely becomes a means, through the armaments industry, to provide people with decent economic conditions at home; the glory of military conquest is altogether absent. Shaw protested in the Preface to this play that his earlier use of the single word 'Superman' had given rise to the false assumption 'that I look for the salvation of society to the despotism of a single Napoleonic Superman, in spite of my careful demonstration of the folly of that outworn infatuation' (*CPP* iii. 20–1). In *Androcles and the Lion* (1912), the Roman emperor is a nobody and the military captain is much less impressive than the Christian Lavinia, while in *The Inca of Perusalem*, 'An Almost Historical Comedietta' written in 1915, the Kaiser-figure, a distinctly unheroic war leader, says that he considers his ancestor Bedrock the Great a grossly overrated monarch (*CPP* iv. 975) —which we could take as another disrespectful gesture in Carlyle's direction.

It is not altogether surprising, then, to find that Shaw, in the Preface to *Back to Methuselah* (1921), put the military hero in with some unsavoury company, in his complaint that 'Our schools teach the morality of feudalism corrupted by commercialism, and hold up the military conqueror, the robber baron, and the profiteer, as models of the illustrious and the successful' (*CPP* v. 262). It was Shaw's belief—his late belief, at any rate—that it is the incompetence of normal governments which gives the military Great Man his chance. If the Bourbons had governed efficiently, then the French Revolution would have been averted, 'and its reforms carried out under Louis XVI quite constitutionally, in which case Napoleon would have been Heaven knows what: perhaps discharged from the army for outstaying his leave, perhaps an elderly marshal in the royal service, which was his own guess. It does not matter now; but the lesson of his career and that of his twentieth-century imitators remains: to wit, that incompetent governments with obsolete ideologies, however democratic they may be in form, go down before up-to-date conquerors.' This comes from the end of a chapter in *Everybody's Political What's What?* headed 'The Military Man', and in a later chapter, 'Government by Great Men, So-Called', we are told specifically that 'All the good that Great Men have done could have been done without them had the governments they supplanted been either efficient or reasonable.' This chapter outlines various defects of government by Great Men, and states: 'There is no hope for civilization in government by idolized single individuals. Councils of tested qualified persons, subject to the sternest possible public criticism, and to periodical (in pressing cases even summary) removal and replacement, is our safest aim.'[18] I do not want to stray into a general discussion of Shaw's political opinions here, but I think it is clear that any notion of his uncritical hero-worship must be subject to modification.

And yet Shaw did have some laudatory comments to make about Mussolini, and even Hitler at times, and he regarded himself as a consistent supporter of Stalin. To some extent, this reflects his delight in assuming unpopular minority positions in order to attract attention to his utterances and to make his readers think. In the case of Stalin, though, Shaw believed he had found a Great Man who transcended the limitations of such figures as Peter the Great, Napoleon, Kemal Ataturk, Mussolini, and Hitler. '[O]nly Cromwell with his Bible and Covenants of Grace, and Stalin with his Marxist philosophy, held themselves within constitutional limits (as we say, had any principles); and they alone stand out as successful rulers.'[19] Shaw also made a distinction between the military conqueror and the strong national, domestic leader, and while his admiration for the former declined over the years, his admiration for the latter grew at times as a reaction against democratic incompetence. The attitude of Old Hipney, the disillusioned working-class Radical in *On the Rocks* (1933), has at least some of the authority of the play behind it: 'I'm for any Napoleon or Mussolini or Lenin or Chavender that has the stuff in him to take both the people and the spoilers and oppressors by the scruffs of their silly necks and just sling them into the way they should go with as many kicks as may be needful to make a thorough job of it' (*CPP* vi. 719). Sir Arthur Chavender, the ineffectual prime minister in the play, recognizes

---

[18] *Everybody's Political What's What?*, 137–8, 340–1.
[19] *Everybody's Political What's What?*, 339–40. Shaw, during a meeting with Stalin in Moscow in 1931, asked him whether he had ever heard of Cromwell and his precept to put your trust in God and keep your powder dry ('Biographers' Blunders Corrected', *Sixteen Self Sketches* (London: Constable, 1949), 85).

that the country needs men of action rather than mere talkers, and the play as a whole indicates the need for strong political leadership in order to prevent the country from going (in Carlyle's phrase) on the rocks. In English history, this was the realization of Cromwell, who tried to govern constitutionally with a Parliament but 'was finally driven to lay violent hands on parliament, and rule by armed force'.[20]

There are times in history when autocratic government is necessary: that is, when the normal structures of government have become incapable of dealing with the problems of the day. Thus Cromwell's methods were the appropriate ones in the 1650s. One must be careful not to simplify Shaw's views, however. It is easy to see him as the advocate of despotic government, if one is selective in looking at what he wrote. But one of his major works on the art of government is his late play, 'In Good King Charles's Golden Days', and when we look at this work we discover that differing historical circumstances demand different types of government. Throughout his writings Shaw is certainly sympathetic to Cromwell, and to Cromwell as a despot, but in this play we have a highly sympathetic representation of Charles II. Shaw's Charles is like Cromwell in that as a ruler he grasps the political realities of his age, but the political realities have changed by 1680 (the date of the play), and in fact they had changed between the last years of Cromwell's life in the 1650s and the Restoration in 1660. Charles, like King Magnus in *The Apple Cart* (1928), is a wise man and an able monarch rather than a powerful Carlylean hero, and it is instructive to compare the respective central figures in Shaw's first history play and his last—*The Man of Destiny* and *Good King Charles*. Charles is not a man of insatiable, indomitable will, and he is not an initiating, critical force in history. Instead, he makes the best of his circumstances and tries to prevent other people's fanaticism from wrecking the country. In *Everybody's Political What's What?*, there is a reference to 'the wisest of the French kings, grandfather to the wisest king of England',[21] which means Henri IV and Charles II. Although Carlyle insists on wisdom as one of the

---

[20] Shaw, *The Intelligent Woman's Guide to Socialism, Capitalism, Sovietism and Fascism* (London: Constable, 1949), 317.
[21] *Everybody's Political What's What?*, 134.

characteristics of the political leader, Shaw's Charles II is decidedly not a Carlylean hero, and, as we have seen, Carlyle despised Charles II and the Restoration period. In moving from Napoleon to Charles II as the central figure in a history play, Shaw is in one sense moving away from Carlyle.

But Charles II is not the only historical figure celebrated by the later Shaw. The later (let us say post-First-World-War) Shaw rejects the Napoleonic military conqueror as hero, and offers two alternatives. One is the type of leader represented by Charles II or King Magnus, the leader who practises sensible statesmanship. The other is the inspired prophet. In *Good King Charles*, Charles himself is not the only impressive figure in the play; there are at least three others: the artist Kneller, the scientist and philosopher Newton, and the religious prophet George Fox. Fox is not a political leader, but in Shaw's best-known history play we have the person who sees visions as the most significant political force of her time and place. Whereas Napoleon in *The Man of Destiny* is possessed by a devouring devil inside him who must be fed with action and victory (*CPP* i. 650), Joan is possessed by a religious idea. She is not merely the military conqueror. For her, conquest is not an end in itself, but a means towards the fulfilment of a religious purpose. Thus in the movement from Shaw's early Napoleon play to his mature treatment of Joan of Arc, we can see a growing assertion of the importance of the idea in history—the idea as opposed to the mere will to power. In this sense Shaw has moved closer to Carlyle, who wrote in his essay on Sir Walter Scott that 'A great man is ever . . . possessed with an *idea*' (the example that Carlyle gives here, incidentally, is Napoleon).[22]

One might think that Shaw's Julius Caesar anticipates Joan more than his Napoleon does, but even Caesar, although he has ideas, is not possessed by any of them to the extent that his career in *Caesar and Cleopatra* is the expression of one. Joan is effective in the way that Shaw believed Cromwell and Stalin were effective: they were carrying a conception of the world into the realm of action. Joan inspires those around her by conveying her idea to them, whereas Caesar's ideas make no converts in *Caesar and Cleopatra*; we may compare Caesar's cynical follower Rufio with

---

[22] Carlyle, 'Sir Walter Scott', *Critical and Miscellaneous Essays*, iv. 37, *Works*, xxix.

Joan's inspired follower Dunois. The nature of Joan's power is described by Don Juan in the Hell Scene of *Man and Superman*:

> DON JUAN. ... [M]en never really overcome fear until they imagine they are fighting to further a universal purpose—fighting for an idea, as they call it. Why was the Crusader braver than the pirate? Because he fought, not for himself, but for the Cross. What force was it that met him with a valor as reckless as his own? The force of men who fought, not for themselves, but for Islam. They took Spain from us though we were fighting for our very hearths and homes; but when we, too, fought for that mighty idea, a Catholic Church, we swept them back to Africa.
>
> (Act III, *CPP* ii. 657.)

Shaw's historical Great Men in his plays of the 1890s, Napoleon and Caesar, could come under the heading of Carlyle's last lecture in *Heroes and Hero-Worship*: 'The Hero as King', in which Carlyle's two examples are Napoleon and Cromwell. Shaw's later hero, Saint Joan, could be placed under two of Carlyle's other headings. Carlyle's fourth lecture, 'The Hero as Priest', discusses Luther and Knox, and the parallel with Joan as Protestant is obvious. The second lecture is entitled 'The Hero as Prophet', and it concerns itself with only one figure: Mahomet. As we have already seen, Shaw was planning in 1909 to write a play about Mahomet, and a character in *Getting Married* (1907–8) is expressing Shaw's own point of view when he says, 'The character of Mahomet is congenial to me. I admire him, and share his views of life to a considerable extent' (*CPP* iii. 658). When Shaw told the Parliamentary Committee on Stage Censorship in 1909 that 'Great religious leaders are more interesting and more important subjects for the dramatist than great conquerors' (Preface to *The Shewing-Up of Blanco Posnet*, *CPP* iii. 714), he was, in his writing career, precisely midway between his Napoleon and his Joan: it had been fourteen years since he wrote *The Man of Destiny* and it was fourteen years later that he wrote *Saint Joan*. The dramatic subject that Shaw was telling the Parliamentary Committee about, we will recall, was Mahomet.

In 1933 Shaw described Mahomet as 'a great Protestant religious force like George Fox or Wesley',[23] and he was well aware of the connection between Mahomet and Joan. In the Preface to

---

[23] Shaw to Ensor Walters, in *Everybody's Political What's What?*, 228.

Saint Joan he compared them as conqueror-saints, and said that Joan, 'like Mahomet, was always ready with a private revelation from God to settle every question and fit every occasion', and he noted that both 'wrote letters to kings calling on them to make millennial rearrangements' (*CPP* vi. 18, 48, 38). In the fourth scene of the play itself Cauchon makes the comparison between them—not surprisingly, since the Crusades are part of the play's historical background. The Devil, he tells Warwick, is spreading a new heresy everywhere, and he gives Wyclif and Hus as examples. Then he adds a third example:

CAUCHON. ... By [this heresy] an Arab camel driver drove Christ and His Church out of Jerusalem, and ravaged his way west like a wild beast until at last there stood only the Pyrenees and God's mercy between France and damnation. Yet what did the camel driver do at the beginning more than this shepherd girl is doing? He had his voices from the angel Gabriel: *she* has her voices from St Catherine and St Margaret and the Blessed Michael. He declared himself the messenger of God, and wrote in God's name to the kings of the earth. Her letters to them are going forth daily.

(*CPP* vi. 134–5.)

The connection between Joan and Mahomet is also evident if we look at Carlyle's essay on the Prophet in *Heroes and Hero-Worship*. The whole of Carlyle's lecture can be read with Shaw's *Saint Joan* in mind. Carlyle's Mahomet, like Shaw's Joan, is fired with an unshakeable conviction of the truth, of divine revelation. 'No *Dilettantism* in this Mahomet; it is a business of Reprobation and Salvation with him, of Time and Eternity: he is in deadly earnest about it!' Elsewhere in this lecture Carlyle observed in relation to Mahomet that 'Every new opinion, at its starting, is precisely in a *minority of one*. In one man's head alone, there it dwells as yet. One man alone of the whole world believes it; there is one man against all men.'[24] This applies to Joan's sense that the French must obey God's will and expel the English, and it has an application to much of Shaw's other writing as well.

One is often reminded of *Saint Joan* in Carlyle's writings. Carlyle presented Cromwell as a man of intense religious energy, a Protestant hero with an unshakeable conception of what is right and at the same time a shrewd political leader with great common

---

[24] Carlyle, *Heroes and Hero-Worship, Works*, v. 73, 61 ('The Hero as Prophet').

sense. In beginning the section on Cromwell's war with Scotland in *Oliver Cromwell's Letters and Speeches*, Carlyle discussed the weaknesses of the Scots, in a way that suggests the extent to which Joan is a Carlylean hero. The great fault of the Scottish people is that they have produced no sufficiently heroic man among them: 'No man that has an eye to see beyond the letter and the rubric; to discern, across many consecrated rubrics of the Past, the inarticulate divineness too of the Present and the Future, and dare all perils in the faith of that!' Instead of being governed by a hero (as England was at this time) they were governed by uninspired pedants, and 'there is no creature more fatal than your Pedant; safe as he esteems himself, the terriblest issues spring from him. Human crimes are many: but the crime of being deaf to the God's Voice, of being blind to all but parchments and antiquarian rubrics when the Divine Handwriting is abroad on the sky, —certainly there is no crime which the Supreme Powers do more terribly avenge!'[25] Passages like these convey nothing of the element of comedy in *Saint Joan*, but they do convey something of the nature of heroism in the play and the forces of resistance which this heroism must encounter.

In his early *Life of Schiller*, Carlyle wrote about Joan herself in his discussion of Schiller's *Die Jungfrau von Orleans*. Like Shaw in his Preface to *Saint Joan* he dismissed the distortions in previous treatments of Joan by Shakespeare and Voltaire, and he examined the qualities that made Joan great and powerful.

This peasant girl, who felt within her such fiery vehemence of resolution, that she could subdue the minds of kings and captains to her will, and lead armies on to battle, conquering, till her country was cleared of its invaders, must evidently have possessed the elements of a majestic character. Benevolent feelings, sublime ideas, and above all an overpowering will, are here indubitably marked. . . . The strength of her impulses persuades her that she is called from on high to deliver her native France; the intensity of her own faith persuades others; she goes forth on her mission; all bends to the fiery vehemence of her will; she is inspired because she thinks herself so.[26]

---

[25] Carlyle, *Oliver Cromwell's Letters and Speeches*, ii. 169, 171, *Works*, vii (Part VI).
[26] Carlyle, *The Life of Friedrich Schiller*, *Works*, xxv. 155, 157.

Leaving aside Schiller's *Die Jungfrau von Orleans* itself, we can observe a close relationship between Shaw's play and Schiller's as Carlyle sees it. In particular, there is Carlyle's emphasis on the strength of Joan's will. The issue for Carlyle is not whether Joan has received literal divine inspiration from without, but rather the force of her own mind. For Carlyle, as for Shaw, the human mind and will are the real miracle of the universe. In Carlyle's words, Joan's is the triumph 'of Mind over Fate, of human volition over material necessity'.[27]

It is also possible to be reminded of Shaw's Joan in another work of Carlyle's: *The French Revolution*. Joan herself, one could say, led a French revolution in the late Middle Ages, and, much more specifically, there is Carlyle's description of Danton's speech to the Legislative Committee of General Defence in September 1792, during the Allied war against France. 'Strong is that grim Son of France and Son of Earth; a Reality and not a Formula he too.'

'Legislators!' so speaks the stentor-voice, as the Newspapers yet preserve it for us, 'it is not the alarm-cannon that you hear: it is the *pas-de-charge* against our enemies. To conquer them, to hurl them back, what do we require? *Il nous faut de l'audace, et encore de l'audace, et toujours de l'audace*, To dare, and again to dare, and without end to dare!'—Right so, thou brawny Titan; there is nothing left for thee but that. Old men, who heard it, will still tell you how the reverberating voice made all hearts swell, in that moment; and braced them to the sticking-place; and thrilled abroad over France, like electric virtue, as a word spoken in season.[28]

And what is it that Shaw's Joan says when France needed to rise to its own defence three and a half centuries earlier? 'In His strength I will dare, and dare, and dare, until I die' (Scene v, *CPP* vi. 154).[29] I am not suggesting any necessary direct influence here; the line is a well-known one. But Carlyle's passage about Danton does indicate a Carlylean quality in *Saint Joan*. With Joan's

---

[27] Carlyle, *Life of Schiller*, 157.
[28] Carlyle, *The French Revolution*, iii. 23–4, *Works*, iv (Part III, Book i, chap. IV).
[29] Cf. Joan's earlier declaration to the Dauphin in Scene ii: 'I shall dare, dare, and dare again, in God's name, (*CPP* vi. 115); and also General Mitchener's speech in Shaw's 1909 playlet about the Suffragettes, *Press Cuttings*: 'What is life but daring, man? "To dare, to dare, and again to dare..."' (*CPP* iii. 846). Here, however, the daring is just foolish bravado.

peasant background, sense of divine inspiration, and power of will, she is the most Carlylean hero in Shaw's plays.

There are two main respects in which Joan differs from Carlyle's conception of the hero. First, in Carlyle's universe, there are certain fundamental, unchanging truths. When mankind loses touch with them, it is the function of the hero to re-establish the lost contact, to demolish the shams in order to bring people face to face with reality. 'It is the property of every Hero, in every time, in every place and situation, that he come back to reality; that he stand upon things, and not shows of things,' Carlyle said in his lecture on Luther in *Heroes and Hero-Worship*. 'There is nothing generically new or peculiar in the Reformation; it was a return to Truth and Reality in opposition to Falsehood and Semblance, as all kinds of Improvement and genuine Teaching are and have been.'[30] In *Saint Joan* there is emphasis on the newness, the originality of what Joan represents. Her Protestantism is not only new to her contemporaries; it is a new stage in human development. She is not so much coming back to reality as pressing forwards towards a new reality.

Second, Carlyle stresses the decisive role of the individual in history rather more than Shaw does. Carlyle does acknowledge the importance of the age, and he does say that 'Not a Hero only is needed, but a world fit for him; a world not of *Valets*;—the Hero comes almost in vain to it otherwise!'[31] This would correspond to Joan's cry at the end of Shaw's Epilogue, asking when the world will be ready to receive God's saints (*CPP* vi. 208). There are also metaphors in *The French Revolution* suggesting the inevitability of historical development, but in the same work we read that 'had Mirabeau lived, the History of France and of the World had been different', and in the lecture on Luther in *Heroes and Hero-Worship*, that at the Diet of Worms 'had Luther in that moment done other, it had all been otherwise',—the word 'all' taking in subsequent European and American history.[32] *Saint Joan* does not make one feel that if Joan had lived, the history of France or the world would have been different, or that had she done otherwise, 'it had all been otherwise'. Joan is part of historical

---

[30] Carlyle, *Heroes and Hero-Worship*, *Works*, v. 123, 124 ('The Hero as Priest').
[31] Carlyle, *Heroes and Hero-Worship*, *Works*, v. 216 ('The Hero as King').
[32] Carlyle, *The French Revolution*, ii, *Works*, iii. 138 (Part II, Book iii, chap. VI); *Heroes and Hero-Worship*, *Works*, v. 135 ('The Hero as Priest').

forces that are wider than herself; she is the incarnation of the emerging *zeitgeist*, not its sole cause. Her mind is part of the intellectual current of the age, and she is not in a minority of one to quite the same extent that Carlyle's Mahomet is. In the second scene of *Saint Joan*, the Archbishop, before Joan's arrival at the Court in Chinon, explains to La Trémouille that 'There is a new spirit rising in men: we are at the dawning of a wider epoch' (*CPP* vi. 107). In the fourth scene both Cauchon and Warwick are aware of the new forces of Protestantism and Nationalism, the latter of which is evident in de Stogumber. Cauchon refers to 'Frenchmen, as the modern fashion calls them' (*CPP* vi. 128), and in the sixth scene Cauchon warns that 'we are confronted today throughout Europe with a heresy that is spreading' among strong-minded people (*CPP* vi. 167–8). Warwick, Cauchon, and the Inquisitor see Joan as representative of already present forces, rather than as an independent cause. She is not trans-historical, standing above history, but is of her time and place, made possible by historical conditions. Her will plays a critical part in historical development, but her actions are possible only within her specific historical context. Carlyle would not have found this picture altogether unacceptable, but his emphasis is decidedly more on the individual and less on the age.

Joan is a World-Historical Individual in Hegel's use of this term, and Hegel's comments on Julius Caesar would apply more accurately to Shaw's Joan than to his Caesar:[33]

> It was not... his private gain merely, but an unconscious impulse that occasioned the accomplishment of that for which the time was ripe. Such are all great historical men—whose own particular aims involve those large issues which are the will of the World-Spirit.... Such individuals had no consciousness of the general Idea they were unfolding, while prosecuting those aims of theirs; on the contrary, they were practical, political men. But at the same time they were thinking men, who had an insight into the requirements of the time—*what was ripe for development*. This was the very Truth for their age, for their world; the species next in order, so to speak, and which was already formed in the womb of time. It was theirs to know this nascent principle; the necessary, directly sequent step in progress, which their world was to take; to make this

---

[33] For a discussion of Shaw's Caesar and Joan as Hegelian World-Historical Individuals, see Robert F. Whitman, *Shaw and the Play of Ideas* (Ithaca, NY: Cornell Univ. Press, 1977), 202–10, 267–74.

their aim, and to expend their energy in promoting it. World-historical men—the Heroes of an epoch—must, therefore, be recognized as its clear-sighted ones; *their* deeds, *their* words are the best of that time.[34]

Joan senses what is ripe for development in her epoch, and Shaw, like Hegel, sees the individual hero in the context of the epoch.

[34] G. W. F. Hegel, *The Philosophy of History* (tr. J. Sibree. New York: Dover, 1956), 30.

3

The Middle Ages, the Renaissance, and After

Shaw told Ellen Terry in 1897 that he required whole populations and historical epochs to engage his interests seriously;[1] and in the case of epochs, as in the case of populations, he took an interest in the distinctions between one and another. In the same way that he was willing to generalize about characteristics of the English, the Irish, the Americans, the Jews, and so on, he was willing to generalize about characteristics of particular historical periods. Something of Shaw's way of approaching a variety of subjects is revealed in his comment, in a musical review in 1889, that 'Progression by semitones is too gradual for my ardent nature.'[2] Although (as we shall see in the next chapter) Shaw had a sharp eye for historical parallels, he often liked to draw clear divisions between individuals, between nations, and between historical periods. Thus, in spite of the deliberate anachronisms in *Saint Joan*, the play ensures that an audience will perceive and understand the medieval setting. There is an element of the history lesson in the play, and one of the dramatist's pedagogical techniques is to make his characters more conscious of their epoch than their historical originals would have been. 'Cauchon and Lemaître have to make intelligible not only themselves but the Church and the Inquisition, just as Warwick has to make the feudal system intelligible,' Shaw wrote in the Preface to *Saint Joan*, 'the three between them having thus to make a twentieth-century audience conscious of an epoch fundamentally different from its own' (*CPP* vi. 73). Each epoch is distinct, and Shaw's idea of history includes the succession of epochs. 'What the common man wants', he said in reviewing G. K. Chesterton's *History of England*, 'is not a history of the kings or the priests, or

[1] Shaw to Ellen Terry, 8 Sept. 1897, *Collected Letters 1874–1897*, ed. Dan H. Laurence (London: Max Reinhardt, 1965), 801.
[2] *Shaw's Music*, ed. Dan H. Laurence (3 vols., London: Max Reinhardt, The Bodley Head, 1981), i. 603.

the nobs, or the snobs, or any other set, smart or slovenly, but a vigorously comprehended and concisely presented history of epochs.'[3]

In Shaw's early Fabian years he may have preached a Marxist conception of epochs as embodying solely economic relationships—a stock of generalizations about the evolution from slavery to serfdom and from serfdom to free wage labour.[4] But notice what it is that Cauchon, Lemaître, and Warwick have to convey to an audience about the Middle Ages. In order to make his play intelligible, Shaw had to endow them with enough consciousness of their epoch 'to enable them to explain their *attitude* to the twentieth century' (*CPP* vi. 74; italics added). What defines an epoch is the attitudes of the people who lived in it; some of these attitudes will be economic, but political, religious, aesthetic, and other considerations are critical as well. Just as Shaw's later conception of the Great Man is a person who incarnates an idea, so his conception of a historical epoch is a period of time that embodies an idea. And just as, in the Epistle Dedicatory to *Man and Superman*, he objected to Shakespeare's heroes as lacking ideas (*CPP* ii. 520), so in the Preface to *Saint Joan* he objected to the false impression that Shakespeare's plays convey about the nature of history: 'His kings are not statesmen: his cardinals have no religion: a novice can read his plays from one end to the other without learning that the world is finally governed by forces expressing themselves in religions and laws which make epochs rather than by vulgarly ambitious individuals who make rows' (*CPP* vi. 70–1).

The context of this depreciation of Shakespeare is a discussion of Shaw's own qualification for writing a play about the Middle Ages. Shaw claimed, as a post nineteenth-century writer, to have an advantage over the Elizabethans. 'I write in full view of the Middle Ages, which may be said to have been rediscovered in the middle of the nineteenth century after an eclipse of about four hundred and fifty years. The Renascence of antique literature and art in the sixteenth century, and the lusty growth of Capitalism, between them buried the Middle Ages; and their resurrection is a second Renascence' (*CPP* vi. 70). Shaw, who revered William

---

[3] Shaw, 'Something Like a History of England', *Pen Portraits and Reviews* (London: Constable, 1949), 90.
[4] See above, pp. 19–20.

Morris as he never revered any other contemporary, regarded himself as part of this nineteenth-century rediscovery of the Middle Ages.[5] Like Morris and Ruskin, he was attracted not only to medieval art but also to the whole civilization that produced it. In his 1896 essay for the *Savoy* magazine, 'On Going to Church', he argued that 'the decay of religious art from the sixteenth century to the nineteenth was not caused by any atrophy of the artistic faculty, but was an eclipse of religion by science and commerce';[6] that is, the aesthetic superiority of the Middle Ages is owing to the medieval world's superiority as a society.

Shaw's fifth play, *Candida*, was subtitled 'A Mystery', to suggest the religious art of the Middle Ages. In the Preface to *Plays Pleasant* he explained the genesis of the work by recounting, 'In the autumn of 1894 I spent a few weeks in Florence, where I occupied myself with the religious art of the Middle Ages and its destruction by the Renascence.' On his return he felt that 'the time was ripe for a modern pre-Raphaelite play' (*CPP* i. 372), and in the first act of *Candida* we have something like a conflict between medieval and Renaissance values in the encounter between the Christian Socialist clergyman Morell and his Capitalist father-in-law Burgess. At the end of the play the poet Marchbanks asserts the primacy of the religious impulse by flying out into the night to seek a higher destiny than mere domestic contentment. *Candida* is not really much of a medieval play, or Pre-Raphaelite play, but parts of it, and more especially parts of the Preface to *Plays Pleasant*, suggest Shaw's attraction to the Medievalist movement of the nineteenth century.

Another work of the 1890s that suggests this attraction is *The Perfect Wagnerite*, which Shaw described in a letter to Sidney Webb as 'an explanatory pamphlet ... on Der Ring des Nibelungen, jam full of Socialism in the manner of Ruskin'.[7] Shaw interprets *The Ring* as an allegory about the loss of a golden age; and he reads Wagner with the eyes of Ruskin and Morris in

---

[5] Shaw's affiliations with Morris, Ruskin, and Pre-Raphaelitism are examined in Elsie B. Adams, *Bernard Shaw and the Aesthetes* ([Columbus]: Ohio State Univ. Press, 1971), 3–30.

[6] Shaw, 'On Going to Church', in *Selected Non-Dramatic Writings of Bernard Shaw*, ed. Dan H. Laurence (Cambridge, Mass.: Riverside, Houghton Mifflin, 1965), 383.

[7] Shaw to Sidney Webb, 7 May 1898, *Collected Letters 1898–1910*, ed. Dan H. Laurence (London: Max Reinhardt, 1972), 41.

hinting that Wotan and the gods represent medieval society which is destroyed by Renaissance commercialism. The gods 'may go to these honest giants who will give a day's work for a day's pay, and induce them to build for Godhead a mighty fortress, complete with hall and chapel, tower and bell, for the sake of the homesteads that will grow up in security round that church-castle. This only, however, whilst the golden age lasts. The moment the Plutonic power is let loose, and the loveless Alberic comes into the field with his corrupting millions, the gods are face to face with destruction.'[8]

Shaw's attraction to the Middle Ages often expresses itself in the way that Carlyle's did, in the use of the period to demonstrate the relative iniquity or ugliness of his own time. Thus in the *Intelligent Woman's Guide to Socialism* he drew a contrast between the feudal aristocracy in the Middle Ages and the plutocracy of the present day. Medieval landowners, and the king himself, had real responsibilities, while their modern counterparts have nothing to do but amuse themselves while other people do the work. 'Henry IV, who died of overwork, found to his cost how true it was in those days that the greatest among us must be servant to all the rest. Nowadays it is the other way about: the greatest is she to whom all the rest are servants.'[9]

The 1907–8 discussion play *Getting Married* is set in the present in the Palace of the Bishop of Chelsea, who refers to William Morris (without actually naming him) as 'a great English poet of my acquaintance' (*CPP* iii. 577). The opening stage direction tells us that the Bishop appreciates the Norman building in which he lives, and the passage contrasts medieval building methods with modern ones. The house is built to last forever, as if its builders '*had resolved to shew how much material they could lavish on a house built for the glory of God, instead of keeping a competitive eye on the advantage of sending in the lowest tender, and scientifically calculating how little material would be enough to prevent the whole affair from tumbling down by its own weight*' (*CPP* iii. 547). In 1890, in a discussion of a lecture by a French baritone, Shaw also introduced this 'past and present' theme. He referred to the superiority of medieval building, this

[8] Shaw, *The Perfect Wagnerite*, in *Shaw's Music*, iii. 430.
[9] Shaw, *The Intelligent Woman's Guide to Socialism, Capitalism, Sovietism and Fascism* (London: Constable, 1949), 166.

time contrasting medieval and modern taste. 'I have no fault to find with the lecture, except on a point of history,' he wrote. 'All that about the Dark Ages and the barbarous Middle Ages is a modern hallucination, partly pious, partly commercial. There never were any Dark Ages, except in the imagination of the Blind Ages. Look at their cathedrals and their houses; and then believe, if you can, that they were less artistic than we who have achieved the terminus at Euston and the Gambetta monument.'[10]

Shaw's finest tribute to William Morris was the obituary article which he wrote in the *Saturday Review* in 1896, in which (in one passage) he defended him against the charge of being archaic. Morris, argued Shaw, was far ahead of his time in his designs and in his narratives: '[H]is hangings, his tapestries, and his printed books have the twentieth century in every touch of them; whilst as to his prose word-weaving, our worn-out nineteenth-century Macaulayese is rancid by comparison. He started from the thirteenth century simply because he wished to start from the most advanced point instead of from the most backward one—say 1850 or thereabout.'[11] Of course, by the time the twentieth century actually arrived, Shaw no longer regarded it as a better future but rather as a present that contrasted unfavourably with the Middle Ages. There is much of this kind of argument in the Preface to *Saint Joan*. In one section of this Preface, on the superstitious scientific education that a twentieth-century Joan would receive, Shaw asked:

Does not the present cry of Back to the Middle Ages, which has been incubating ever since the pre-Raphaelite movement began, mean that it is no longer our Academy pictures that are intolerable, but our credulities that have not the excuse of being superstitions, our cruelties that have not the excuse of barbarism, our persecutions that have not the excuse of religious faith, our shameless substitution of successful swindlers and scoundrels and quacks for saints as objects of worship, and our deafness and blindness to the calls and visions of the inexorable power that made us, and will destroy us if we disregard it? To Joan and her contemporaries we should appear as a drove of Gadarene swine, possessed by all the unclean spirits cast out by the faith and civilization of the Middle Ages, running violently down a steep place into a hell of high explosives.

(*CPP* vi. 31.)

[10] *Shaw's Music*, ii. 230–1.
[11] Shaw, *Our Theatres in the Nineties* (3 vols., London: Constable, 1954), ii. 210.

Joan's story can be told properly only by someone who is sympathetic towards the Middle Ages, and free from the Macaulayish side of nineteenth-century thinking. Her historian 'must understand the Middle Ages, the Roman Catholic Church, and the Holy Roman Empire much more intimately than our Whig historians have ever understood them', and must regard medieval history as 'the record of a high European civilization based on a catholic faith' (Preface to *Saint Joan*, *CPP* vi. 20, 44). Someone who is limited by 'Protestant misunderstandings of the Middle Ages' will never be able to grasp Joan's history, because in order to understand her it is necessary to understand—and to sympathize with—her environment.

To see her in her proper perspective you must understand Christendom and the Catholic Church, the Holy Roman Empire and the Feudal System, as they existed and were understood in the Middle Ages. If you confuse the Middle Ages with the Dark Ages, and are in the habit of ridiculing your aunt for wearing 'medieval clothes,' meaning those in vogue in the eighteen-nineties, and are quite convinced that the world has progressed enormously, both morally and mechanically, since Joan's time, then you will never understand why Joan was burnt, much less feel that you might have voted for burning her yourself if you had been a member of the court that tried her; and until you feel that you know nothing essential about her.

(*CPP*, vi. 43–4.)

I think I have demonstrated the legacy of Victorian Medievalism in Shaw's thinking. Shaw, in the tradition of Carlyle, Ruskin, and Morris, is drawn to medieval civilization, particularly as an alternative to the debased present. Nevertheless, this last sentence that I have quoted must remind us that in Shaw's main dramatic study of the Middle Ages, the forces of medieval civilization burn an innocent and heroic young woman at the stake. Shaw has chosen as his dramatic subject a story that seems to show Medievalism at its worst, to show it in the light of nineteenth-century Protestant, Whig historical attitudes, as barbarous superstition.

Here we come upon the kind of conflict in Shaw's thinking that serves him well as a dramatist. For alongside all the favourable statements about the Middle Ages I have just quoted one could set some unfavourable ones as well. One of the passages we have been looking at occurs in Shaw's exposition of *Das Rheingold* —the evocation of the feudal hall and chapel, tower and bell. Just a few pages later, we are confronted with another side of the Middle Ages: 'In older times, when the Christian laborer was drained dry by the knightly spendthrift, and the spendthrift was drained by the Jewish usurer, Church and State, religion and law, seized on the Jew and drained him as a Christian duty.'[12] The context here suggests that this refers to the later Middle Ages, which is to say the period of *Saint Joan*. In *Everybody's Political What's What?*, we are shown another side of the later Middle Ages in the comment that 'The Black Death suggested an attempt to exterminate the human race in a fit of disgust at the stinking filth of the cities'[13]—which brings us closer to Macaulay's way of looking at the period. In an earlier part of this book Shaw presented an economic history of Europe from the time of William the Conqueror to the present, and here his account of the decline of Feudalism acknowledges that its usefulness came to an end. After a laudatory paragraph about William the Conqueror (one of Carlyle's heroes) and the economic system of land tenure that he established, Shaw observed that 'It was quite a reasonable arrangement under the circumstances, and kept the country in order for a while in an agricultural society of barons, bishops, farmers and serfs. But only for a while. Only as long as the facts corresponded roughly to the plan of feudalism. And facts will not, like plans, stay put.'[14]

Saint Joan was published in the year in which the Dutch historian Johan Huizinga's classic study *The Waning of the Middle Ages* first appeared in English translation.[15] This coincidence is a reminder of an important aspect of Shaw's play. One of its main subjects is the encounter between the late Middle Ages and the

[12] *The Perfect Wagnerite*, in *Shaw's Music*, iii. 437.
[13] Shaw, *Everybody's Political What's What?* (London: Constable, 1944), 220.
[14] *Everybody's Political What's What?*, 9–10.
[15] In 1925 Huizinga wrote an essay on Shaw's *Saint Joan*, in which he commented that 'Shaw would like us to consider [Joan's] trial as nothing more than the necessary defense of her age against the unknown and immeasurable danger that would destroy that age.' Huizinga, however, does not see the historical Joan as a Protestant or a Nationalist, and he argues that her own personality—rather than her place in intellectual history—is the proper object of attention for the historian. 'In her irreducible uniqueness she can be understood only by means of a sense of sympathetic admiration. She does not lend herself to being used to clarify currents and concepts of her day.' That is why Huizinga did not deal with her in *The Waning of the Middle Ages* (which was originally published in Dutch in 1919). ('Bernard Shaw's Saint', *Men and Ideas: History, the Middle Ages, the Renaissance*, tr. James S. Holmes and Hans van Marle (New York: Meridian, 1959), 207–39; see esp. 231, 236–9.)

beginnings of the Renaissance. Shaw claimed in the Preface to have taken care to let the medieval atmosphere blow through his play freely (*CPP* vi. 71), and he told Sydney Cockerell that 'Ruskin and Morris and all the painters were on the job before me' in working over the historical background of the play.[16] In a newspaper interview he went so far as to describe his play as 'a dialogue dealing with religion and with the politics of the Middle Ages'.[17] The Middle Ages in the play, however, are the old epoch that is on the verge of giving way to the Renaissance and modernity. The encounter between one historical epoch and its successor is evident in Joan's first appearance on the stage:

JOAN [*bobbing a curtsey*]. Good morning, captain squire. Captain: you are to give me a horse and armor and some soldiers, and send me to the Dauphin. Those are your orders from my Lord.

ROBERT [*outraged*]. Orders from *your* lord. And who the devil may your lord be? Go back to him, and tell him that I am neither duke nor peer at his orders: I am squire of Baudricourt; and I take no orders except from the king.

JOAN [*reassuringly*]. Yes, squire: that is all right. My Lord is the King of Heaven.

ROBERT. Why, the girl's mad.

(Scene i, *CPP* vi. 85.)

In the distinction between Joan's sense of 'my Lord' and Baudricourt's we see the clash between the Middle Ages and the Renaissance, between the declining epoch and the rising one. As E. M. W. Tillyard says in *Shakespeare's History Plays* about Richard and Bolingbroke in *Richard II*, 'We have in fact the contrast not only of two characters but of two ways of life',[18] and Shakespeare's Richard is the medieval king supplanted by the

[16] Shaw to Sydney Cockerell, 27 Feb. 1924, *Collected Letters 1911–1925*, ed. Dan H. Laurence (London: Max Reinhardt, 1985), 867. Joan's sentimental speech in the Trial scene ('... the wind in the trees, the larks in the sunshine, the young lambs crying through the healthy frost ...') could owe something to William Morris; see *A Dream of John Ball*, chap. x, in which the medieval religious revolutionary figure, disclaiming a fear of death, recalls the time when he and his sister played together while 'the sparrow-hawk wheeled and turned over the hedges and the weasel ran across the path, and the sound of the sheep-bells came to us from the downs as we sat happy on the grass ...' (*William Morris*, ed. G. D. H. Cole (London: Nonesuch, 1944), 248).

[17] James Graham, 'Shaw on *Saint Joan*', *New York Times*, 13 Apr. 1924, Sect. 8, p. 2; reprinted in Stanley Weintraub, ed., *'Saint Joan' Fifty Years After* (Baton Rouge: Lousiana State Univ. Press, 1973), 18.

[18] E. M. W. Tillyard, *Shakespeare's History Plays* (1944; London: Chatto and Windus, 1964), 258.

Middle Ages, Renaissance, and After 85

more modern Bolingbroke thirty years before the time of *Saint Joan*.

In the conflict between the Middle Ages and the Renaissance the play and its author are on both sides at once. Everyone has noticed the play's effort to be fair to the Church, and this is not simply a dramatic technique, but an expression of Shaw's historical values and judgements. Some of the dramatic strength of *Saint Joan* derives from the collision in Shaw's own mind between the Feudal, Catholic values of the Middle Ages and the Capitalist, Protestant values of the Renaissance—between the collective will of a society acting in concert and the energetic individual will that rebels against social authority. Or to put this in more biographical terms, one could say that the play reflects a conflict between Shaw's Protestant background and his exposure to the Medievalist movement of Victorian England. When the man who admired William Morris and described *Candida* as a Pre-Raphaelite play came to write a major play about the Middle Ages, he chose as his subject the breaking up of the era. Whereas the Preface, with its approving talk about the cry of 'Back to the Middle Ages', is written in the William Morris tradition, the play itself celebrates the Renaissance forces of individualism, Protestantism, and Nationalism, and dramatizes the waning of the Middle Ages as a necessary, progressive, and beneficial historical development. And the conflict is not only between the Preface and the play, but within the play itself. The play has something of the quality that Matthew Arnold attributed to the disinterested criticism which recognizes the defects of Protestantism but 'will not on that account forget the achievements of Protestantism in the practical and moral sphere; nor that, even in the intellectual sphere, Protestantism, though in a blind and stumbling manner, carried forward the Renascence, while Catholicism threw itself violently across its path'.[19] Joan's undeclared, unconscious mission is to

[19] Matthew Arnold, 'The Function of Criticism at the Present Time', *Lectures and Essays in Criticism*, ed. R. H. Super (Ann Arbor: Univ. of Michigan Press, 1962), 281. Although Macaulay hardly represents the Arnoldian spirit of disinterested criticism, it may be worth noting here that in the first chapter of the *History of England* he judged that Catholicism and Protestantism were each appropriate in their own time: 'Those who hold that the influence of the Church of Rome in the dark ages was, on the whole, beneficial to mankind may yet with perfect consistency regard the Reformation as an inestimable blessing. The leading strings, which preserve and uphold the infant, would impede the fullgrown man' (*The History of England from the Accession of James II*, *The Works of Lord Macaulay*, ed. Lady Trevelyan (8 vols., London: Longmans, Green, 1879), i. 36–7 (chap. 1)).

destroy medieval civilization, and a production of the play should not allow us to see her only as a positive heroine, but also as a heedlessly destructive force confronting a civilization that is of value in spite of its blindness and rigidity. The Middle Ages must die, but the epoch has its dignity and beauty—a beauty that should be expressed powerfully in the sets and costumes, to remind us of the value of what is being lost.[20]

'I really cannot keep my temper over the Elizabethan dramatists and the Renaissance', Shaw declared in writing about *Macbeth* in 1895. In the early seventeenth century, 'every art was corrupted to the marrow by the orgie called the Renaissance, which was nothing but the vulgar exploitation in the artistic professions of the territory won by the Protestant movement'. This sounds like the voice of the Victorian Medievalist, but the next sentence begins with a reference to Protestantism as 'that great self-assertion of the growing spirit of man',[21] and once again we can see both sides of Shaw's historical outlook. In *Saint Joan* the emphasis is on Protestantism as that great self-assertion of the growing spirit of man, rather than on the Renaissance as an orgy of vulgar artistic exploitation. In considering the play, we are less likely to think of Shaw's connection with William Morris and more likely to think of his background as he described it in the Preface to *John Bull's Other Island*: 'I am a genuine typical Irishman of the Danish, Norman, Cromwellian, and (of course) Scotch invasions. I am violently and arrogantly Protestant by family tradition' (*CPP* ii. 811). We might also bear in mind the subtitle that Shaw suggested to his publisher for *The Perfect Wagnerite* in 1898, 'the New Protestantism',[22] and consider the chapter entitled 'Siegfried as Protestant', parts of which read like a description of Joan in the play. Joan's question at her trial, 'What other judgment can I judge by but my own?' (*CPP* vi. 175), is a declaration of Protestant revolt against the fundamental

---

[20] In the first London production, at the New Theatre in 1924, Charles Ricketts's sets and costume designs were of this kind, and they drew much favourable comment at the time. One contemporary commentator was Huizinga, who wrote, 'I must admit that I have never seen more convincing historical staging than this work of Ricketts's'; and he saw Ricketts 'following in the footsteps of William Morris and Rossetti and Ford Madox Brown' (*Men and Ideas*, 214–15).
[21] Shaw, *Our Theatres in the Nineties*, i. 130–1.
[22] Shaw to Grant Richards, 20 Aug. 1898, *Collected Letters 1898–1910*, 58.

assumptions of the Middle Ages. Buckle called the Reformation 'neither more nor less than open rebellion': 'To establish the right of private judgment, was to appeal from the church to individuals; it was to increase the play of each man's intellect; it was to test the opinions of the priesthood by the opinions of laymen; it was, in fact, a rising of the scholars against their teachers, of the ruled against their rulers.'[23] This is the spirit that Joan represents, and Shaw in selecting her as a Protestant heroine seems to be allying himself not only with Buckle but with Carlyle, who declared that Luther's speech to the Diet of Worms ('Here stand I; I can do no other: God assist me!') was 'the greatest moment in the Modern History of Men'.[24]

On the other hand, Joan, as the incarnation of the Renaissance spirit, represents more in the play than the beneficial assertion of individual judgement. For one thing, she represents modern warfare. She tells Dunois that the French army cannot rely on horses but needs plenty of artillery (Scene iii, *CPP* vi. 121), and that his medieval art of war is of no use because his knights are no good for real fighting.

> JOAN. . . . War is only a game to them, like tennis and all their other games: they make rules as to what is fair and what is not fair, and heap armor on themselves and on their poor horses to keep out the arrows; and when they fall they cant get up, and have to wait for their squires to come and lift them to arrange about the ransom with the man that has poked them off their horse. Cant you see that all the like of that is gone by and done with? What use is armor against gunpowder? And if it was, do you think men that are fighting for France and for God will stop to bargain about ransoms, as half your knights live by doing? No: they will fight to win; and they will give up their lives out of their own hand into the hand of God when they go into battle, as I do.
>
> (Scene v, *CPP* vi. 149.)

A full response to *Saint Joan* must involve the awareness that it was written just five years after the end of the First World War, and that the War was the greatest transforming and disillusioning experience in Shaw's life. To read or watch the play without

---
[23] H. T. Buckle, *History of Civilization in England* (1857–61; 3 vols., London: Longmans, Green, 1878), ii. 140–1.
[24] Thomas Carlyle, *On Heroes, Hero-Worship, and the Heroic in History*, *The Works of Thomas Carlyle*, ed. H. D. Traill (Centenary Edn., 30 vols., London: Chapman and Hall, 1896–9), v. 135 ('The Hero as Priest').

having the War in one's mind is to miss its ironic perspective, and a production should make an audience feel that Joan is a threat as well as a saint. For the fulfilment of her notions of the new warfare was manifested in the trenches of France and Belgium almost 500 years later.

Joan represents the modern warfare that made the horrors of 1914–18 possible, and also the spirit of Nationalism that led to the First World War. The play links these two post-medieval developments in that Joan's methods of warfare are in the service of a Nationalist objective. She says to Dunois in the Epilogue, when she learns that the English have now been expelled from France: 'And you fought them *my* way, Jack: eh? Not the old way, chaffering for ransoms; but The Maid's way: staking life against death ... nothing counting under God but France free and French' (*CPP* vi. 197–8). The men fight to win because they are fighting for their own nation. In *Geneva* (1936), in which we see national rivalries that will lead to a second world war, we find a different perspective on Nationalist warfare. In this play, the Secretary of the League of Nations declares that 'We need something higher than nationalism: a genuine political and social catholicism. How are you to get that from these patriots, with their national anthems and flags and dreams of war and conquest rubbed into them from their childhood? The organization of nations is the organization of world war' (Act II, *CPP* vii. 83).

In *Saint Joan*, Joan's Nationalism is made to sound appealing, but she is not the only Nationalist on the stage. In the Tent Scene, Cauchon issues a threat to de Stogumber: 'If you dare do what this woman has done—set your country above the holy Catholic Church—you shall go to the fire with her' (*CPP* vi. 132). The dangers of the forces that Joan represents are reflected in De Stogumber's presence in the play. He connects, in the mind of the audience, Joan's Nationalism with the fanaticism that led to and flourished during the War. Shaw wrote in the 1919 Preface to *Heartbreak House* that during the War 'There was a frivolous exultation in death for its own sake, which was at bottom an inability to realize that the deaths were real deaths and not stage ones' (*CPP* v. 33); this is precisely De Stogumber's attitude towards the burning of Joan, as he himself discovers to his horror near the end of the Trial Scene. Thus the play provides the audience with evidence to support the fears of Cauchon that

Joan's challenge to the established authority of the Middle Ages will lead to universal war:

CAUCHON. ... What will the world be like when The Church's accumulated wisdom and knowledge and experience, its councils of learned, venerable pious men, are thrust into the kennel by every ignorant laborer or dairymaid whom the devil can puff up with the monstrous self-conceit of being directly inspired from heaven? It will be a world of blood, of fury, of devastation, of each man striving for his own hand: in the end a world wrecked back into barbarism. ... Call this side of her heresy Nationalism if you will: I can find you no better name for it. I can only tell you that it is essentially anti-Catholic and anti-Christian; for the Catholic Church knows only one realm, and that is the realm of Christ's kingdom. Divide that kingdom into nations, and you dethrone Christ. Dethrone Christ, and who will stand between our throats and the sword? The world will perish in a welter of war.

(Scene iv, *CPP* vi. 135, 139–40.)

Joan's Nationalism, which is inseparable from her Protestant individualism, carries the threat of international anarchy.

Joan's Renaissance individualist values have also led to another modern iniquity: Capitalism. She asserts her individual will against the communal authority of the day, and Cauchon predicts that such rebellion will lead to a world 'of each man striving for his own hand' (Scene iv, *CPP* vi. 135). The play does not anachronistically place Joan's individualism in a modern economic context,[25] but a thoughtful, educated, twentieth-century audience might be expected to do so. Shaw himself makes such a connection in discussing the case of Galileo in *Everybody's Political What's What?*, where he sounds much closer to *Saint Joan's* Inquisitor than to the Maid herself. The case, he argues, 'is usually written of as a persecution of a great observer and fearless reasoner by a pack of superstitious, narrow-minded, ignorant

---

[25] R. H. Tawney's *Religion and the Rise of Capitalism*, which began as a series of lectures at King's College, London, in the spring of 1922, a year before Shaw wrote *Saint Joan*, makes the point that early Protestantism had no intention of transforming the medieval economic order: 'If it is true that the Reformation released forces which were to act as a solvent of the traditional attitude of religious thought to social and economic issues, it did so without design, and against the intention of most reformers' ((New York: Harcourt, Brace, 1926), 84). See G. Couchman, *This Our Caesar: A Study of Bernard Shaw's 'Caesar and Cleopatra'* (Mouton: The Hague, 1973), 114–15, on the relationship between Tawney's book and Shaw's Caesar.

priests. This is mere Protestant scurrility.' The Church had to govern simple illiterate people, and 'If Galileo were to tell the people that Joshua should have stopped the earth instead of the sun, and that the story must have been invented by somebody so unlike God as to be grossly ignorant of astronomy, their faith would be shattered; and Christendom would collapse in an orgy of selfish lawlessness.' The Church, it turned out, was quite right, as Galileo himself acknowledged. 'Sure enough, when the truth got about, what the priests feared did largely happen. The quest of salvation gave way to the quest of commercial profits; and Manchester supplanted Rome as the headquarters of civilization.' Elsewhere in this book Shaw wrote of Capitalism as a 'revolution in economic morality, fitting neatly into the great Protestant revolution called The Reformation and The Renascence', and he observed that the Wars of the Roses (which began in 1455, the year before Joan's Rehabilitation, the time of the Epilogue) exterminated the old Feudal nobility 'and gave its powers to a new self-ennobled plutocracy'.[26]

If we look at Joan from the point of view of the declining Middle Ages, then she appears as a necessary and progressive force in history. This is the main perspective within *Saint Joan*, but the play is written for a twentieth-century audience, and from our point of view Joan represents the forces that are now the established ones in our society. Joan's independence of spirit, power of imagination, and strength of will are of enduring, timeless value as human qualities, but the play is not only personal but rather mainly historical in its focus, and the historical forces that Joan incarnates on the stage must now be superseded. Joan is the harbinger of the era that is now waning. Protestant individualism, with its economic and international political implications, is in the position now that the Roman Catholic Church and the Feudal nobility occupied in the fifteenth century. Forward-looking people in the twentieth century will look at the Renaissance in the way that Joan unconsciously

[26] *Everybody's Political What's What?*, 193, 168, 79. For a discussion of the ambiguous response to the Renaissance in *Saint Joan*, see Warren Sylvester Smith, *Bishop of Everywhere: Bernard Shaw and the Life Force* (University Park and London: The Pennsylvania State Univ. Press, 1982), 129–32. Smith quotes a passage from the Victorian historian William Stubbs, on the superiority of the Middle Ages over the Renaissance.

Middle Ages, Renaissance, and After   91

looked at the values of the Middle Ages, as the old era that we must move beyond.

Another figure whom Shaw identified with Renaissance individualism is Shakespeare. In the Preface to *The Dark Lady of the Sonnets*, Shaw said one could make a good case for the proposition that 'one of Shakespear's defects is his lack of an intelligent comprehension of feudalism'; his plays lack any conception of civil public business and he personally was involved in the enclosure of common lands. These attitudes, although they are now becoming out of date as we reach the end of the era in which they have dominated, were valid in Shakespeare's own day. The explanation of his anti-feudal stance is 'not a general deficiency in his mind, but the simple fact that in his day what English land needed was individual appropriation and cultivation, and what the English Constitution needed was the incorporation of Whig principles of individual liberty' (*CPP* iv. 299–300). Such is the double-perspective implicit in *Saint Joan*. What was needed in her day was Protestant individualism, but what is needed in our own day is the supersession of these Renaissance values by a neo-feudal collectivism (as we shall see in Chapter 5 when we consider *John Bull's Other Island*).

Shaw's sympathy with both medieval Feudalism and Protestant individualism also affects his attitudes towards English history in the seventeenth century. On the one hand he is sympathetic to Cromwell and Puritanism, but on the other hand he opposes the development of the Capitalism and Whiggism which he recognizes that Cromwell and the Puritans represent. *Three Plays for Puritans*, the title of the volume that includes two of Shaw's history plays (*The Devil's Disciple* and *Caesar and Cleopatra*), uses the word 'Puritan' not only in its sense of asceticism but in its historical context as well, to refer to English Puritanism of the seventeenth century. In the Preface to the volume Shaw said he thought he had always been a Puritan in his attitude towards art, and he supported his explanation with three illustrious seventeenth-century names: 'I am as fond of fine music and handsome building as Milton was, or Cromwell, or Bunyan.' But he would advocate the destruction of cathedrals and organs if he found they were 'becoming the instruments of a systematic idolatry of sensuousness' (*CPP* ii. 27). In *Everybody's Political What's What?*, he compared himself with the seventeenth-century Puritans by saying that if he were an omnipotent despot he would try to ban the sale of 'analgesics, intoxicants, stimulants, tobacco, fish, flesh, and fowl', with the result that his death 'would be followed by a reaction compared to which the one that followed the death of Cromwell would seem trifling'.[27] We have already noted that Shaw wished to write a play about Cromwell, whom he regarded as a commendably strong leader with a sense of religious purpose.

In a 1907 article about *Caesar and Cleopatra*, however, Shaw talked about seventeenth-century Puritanism in another context, as antithetical to the spirit of his play. *Caesar and Cleopatra*, he suggested, resists the Whig reading of history that glorifies the regicides of 1649. Shakespeare's Julius Caesar is 'nothing but the conventional tyrant of the Elizabethan stage adapted to Plutarch's Roundheaded account of him'.

The reason Shakespear belittled him, and that no later English dramatist touched this greatest of all protagonists until I saw my chance and took it, was simply that Shakespear's sympathies were with Plutarch and the Nonconformist Conscience, which he personified as Brutus. From the date of Shakespear's play onward England believed in Brutus with growing hope and earnestness until the assassination in the Capitol was repeated in Whitehall, and Brutus got his chance from Cromwell, who found him hopelessly incapable, and ruled in Caesar's fashion until he died, when the nation sent for Charles II because it was determined to have anybody rather than Brutus. Yet as late as Macaulay and John Morley you find Brutus still the hero and Caesar still the doubtful character. ('Bernard Shaw and the Heroic Actor', *CPP* ii. 309–10.)

Shaw appears to think of *Caesar and Cleopatra* in terms of seventeenth-century English history, and to see it in opposition to the Puritan Revolution—although he makes a distinction between Cromwell the regicide and Cromwell the enlightened despot. And he connects Shakespeare's *Julius Caesar*, a play he actively disliked, with the execution of Charles I and with a Whig tradition that extends until the late nineteenth century, as he did also in his submission to the Parliamentary Committee on Stage Censorship in 1909, where he suggested that Shakespeare's play 'may quite possibly have helped the regicides of 1649 to see themselves, as it certainly helped generations of Whig statesmen

---

[27] *Everybody's Political What's What?*, 342–3.

to see them, in a heroic light' (Preface to *The Shewing-Up of Blanco Posnet*, *CPP* iii. 705). When we examine Shaw's major play about English history of the seventeenth century, we see once again his tendency to take both sides on historical questions. In a discussion of party politics in *The Intelligent Woman's Guide*, there is a passage that suggests a hostile attitude towards the Restoration. When Cromwell died, Shaw wrote, the split in Parliament was so hopeless that 'there was no way out of the mess but to send for the dead King's son and use him, under his father's title, as the figure head of a plutocratic oligarchy exercising all the old kingly powers and greatly extending them'.[28] This is not altogether inconsistent with the reading of the period in '*In Good King Charles's Golden Days*', in that Charles recognizes the mercantile men of the City of London as the real holders of power in the nation. But whereas the passage in *The Intelligent Woman's Guide* concentrates on the unsavoury economic realities of the Restoration, the play concentrates on the robust intellectual life of the period, and on the urbane and level-headed character of the King himself.

The picture of the King is a clear rejection of Macaulay's account of him in the second chapter of *The History of England*, and it is at least as favourable as the one in Hume's supposedly 'Tory' history. Hume's Charles II lacked the integrity and strict principles of Charles I, but 'was happy in a more amiable manner and more popular address'. He 'had not a grain of pride or vanity in his whole composition, but was the most affable, best-bred man alive'.[29] Macaulay allows Charles some personally agreeable characteristics, but the overall impression he leaves us of the King and his administration is a poor one. He draws attention to Charles's laziness, for example: 'He detested business, and would sooner have abdicated his crown than have undergone the trouble of really directing the administration. Such was his aversion to toil, and such his ignorance of affairs, that the very clerks who attended him when he sate in council could not refrain from sneering at his frivolous remarks, and at his childish impatience.' Macaulay also drew attention to the Treaty of Dover, under

---

[28] *Intelligent Woman's Guide*, 345.
[29] David Hume, *The History of England from the Invasion of Julius Caesar to the Revolution in 1688*, ed. Rodney W. Kilcup (Chicago and London: Univ. of Chicago Press, 1975), 297.

which Charles was to receive a large subsidy and other support from Louis XIV in return for adopting the Roman Catholic religion and maintaining a secret military alliance with France. Macaulay described the role played by Louis's unofficial envoy, the 'handsome, licentious, and crafty Frenchwoman' Louise de Kéroualle, and he cautioned his readers not to assign too much of the blame for the treaty to Charles's cabinet (the Cabal) rather than the King himself. 'We must take heed . . . that we do not load their memory with infamy which of right belongs to their master. For the treaty of Dover the King himself is chiefly answerable. He held conferences on it with the French agents: he wrote many letters concerning it with his own hand: he was the person who first suggested the most disgraceful articles which it contained; and he carefully concealed some of those articles from the majority of his Cabinet.'[30]

Macaulay's description of Charles's laziness is decidedly contradicted by the affable but efficient king portrayed in Shaw's play, and Shaw in his Preface defended Charles's conduct over the Treaty of Dover in a passage that appears to be aimed directly at Macaulay. 'It is inferred', Shaw wrote, 'that [Charles] was politically influenced by women, especially by Louise de Kéroualle, who, as an agent of Louis XIV, kept him under the thumb of that Sun of Monarchs as his secret pensioner.' Shaw's defence of Charles is that the English people would not provide Charles with the money he needed in order to run the country:

> Charles, to carry on, had to raise the necessary money somewhere; and as he could not get it from the Protestant people of England he was clever enough to get it from the Catholic king of France; for, though head of the Church of England, he privately ranked Protestants as an upstart vulgar middle-class sect, and the Catholic Church as the authentic original Church of Christ, and the only possible faith for a gentleman. In achieving this he made use of Louise: there is no evidence that she made use of him. To the Whig historians the transaction makes Charles a Quisling in the service of Louis and a traitor to his own country. This is mere Protestant scurrility: the only shady part of it is that Charles, spending the money in the service of England, gave *le Roi Soleil* no value for it.
>
> (*CPP* vii. 207.)

---

[30] Macaulay, *History of England, Works*, i. 133, 164–6 (chap. II).

Within the play, Charles explains to James that when he becomes king he will have to take his money where he can get it. 'French money is as good as English. King Louis gets little enough for it: I take care of that' (Act I, *CPP* vii. 250), and the audience does not feel that Charles is behaving disgracefully in this respect—or in any other respect. *Good King Charles* may seem an odd play from the pen of the author of *Three Plays for Puritans*, if we are thinking of the word 'Puritan' in its seventeenth-century context. *Good King Charles* is decidedly not a play for Puritans, and in fact Shaw could have published it along with *The Apple Cart* in a volume entitled *Two Plays for Cavaliers*.

Good King Charles aims to make us see England in 1680 not as Macaulay saw it, but as Charles himself might have seen it. Although the play follows Macaulay's *History* in its picture of James as a weak, foolish bigot and bully, it does not put Catholicism itself in a bad light, and it does not give James a monopoly on persecution: the anti-Catholic, persecuting Protestants led by Titus Oates are much more of a threat than James is at the time of the play.[31] As in his view of the Middle Ages and the Renaissance, Shaw is committed to neither the Protestant nor the Roman Catholic side. Catholicism is represented not only by James, but also by his brother Charles. Protestantism is represented not only by Titus Oates and the No Popery mob, but also by the founder of the Society of Friends, George Fox.

Fox is a Protestant who fares very badly in Macaulay's *History*, and is one of the heroes of Carlyle's *Sartor Resartus*. Macaulay dwelt on the peculiar alliance between James II and the Quakers (William Penn in particular), and in recording Fox's death in 1691 he devoted several scoffing pages to his eccentricities. For example: 'He long wandered from place to place, teaching this strange theology, shaking like an aspen leaf in his paroxysms of fanatical excitement, forcing his way into churches, which he nicknamed steeple houses, interrupting prayers and sermons with clamour and scurrility, and pestering rectors and justices with epistles much resembling burlesques of those sublime odes in

---

[31] Buckle's *History of Civilization* gives the behaviour of French Protestants in the early seventeenth century, under Louis XIII, as one of 'many instances which show how superficial is the opinion of those speculative writers, who believe that the Protestant religion is necessarily more liberal than the Catholic' (ii. 51, chap. I). See also iii. 275–6, on Scotland in the seventeenth century (chap. IV).

which the Hebrew prophets foretold the calamities of Babylon and Tyre.'[32] To Carlyle, on the other hand (or at least to his Teufelsdröckh),

> Perhaps the most remarkable incident in Modern History . . . is . . . an incident passed carelessly over by most Historians, and treated with some degree of ridicule by others: namely, George Fox's making to himself a suit of Leather. This man, the first of the Quakers, and by trade a Shoemaker, was one of those, to whom, under ruder or purer form, the Divine Idea of the Universe is pleased to manifest itself; and, across all the hulls of Ignorance and earthly Degradation, shine through, in unspeakable Awfulness, unspeakable Beauty, on their souls: who therefore are rightly accounted Prophets, God-possessed; or even Gods, as in some periods it has chanced.[33]

Someone with a taste for puns might say that while Carlyle stresses Fox's honest breeches of leather, Macaulay stresses his awkward breaches of good taste. Shaw, who according to Warren Sylvester Smith had apparently read Fox's *Journal*,[34] stresses both the authenticity and the outlandishness. His Fox is much closer to Carlyle's than to Macaulay's, in that his 'Man in Leather Breeches' is a strong prophetic voice of great impressiveness.

Shaw's attitude towards the reign of William III was also very different from Macaulay's. Although he wrote that because he himself was born in 1856 in Dublin, 'the smoke of battle from the Boyne had not cleared away from my landscape, nor the glorious pious and immortal memory of Dutch William faded from my consciousness, when my sense of history was formed',[35] his interest in William III was largely limited to the rise of the party system, and Shaw in his later years took a very disparaging view indeed of the party system. He never tired of recounting the manner of its inception, describing this event as 'Sunderland's trick'—that is, a device proposed to William by the Earl of Sunderland as a means of controlling the House of Commons in order to pursue the war against Louis XIV. 'It was', Shaw wrote

---

[32] Macaulay, *History of England*, *Works*, iii. 388 (chap. XVII).
[33] Carlyle, *Sartor Resartus*, *Works*, i. 166 (Book iii, chap. 1).
[34] Smith, *Bishop of Everywhere*, 69. 'I know of no specific reference that Shaw makes to Fox's *Journal*, but the picture he draws of Fox could hardly have emerged from any other source, and the words he puts into Fox's mouth are sometimes close to *Journal* quotations.'
[35] Stanley Weintraub, ed., *Shaw: An Autobiography 1856–1898* (New York: Weybright and Talley, 1969), 14 (taken from a 1935 article in *G. K.'s Weekly*).

in 1917, 'by far the most revolutionary act of the glorious, pious, and immortal Dutchman to whom England was nothing but a stick to beat Louis XIV, and who found that without the party system the stick would break in his hand as fast as he could splice it.'[36] In 1934 he told Winston Churchill that the party system 'destroyed parliament as the Revolution destroyed the monarchy, and substituted the anarchy of Capitalism. It led to Dickens's Little Dorrit in the XIX century, and in the XX to what you have lived to see: a disgust with Macaulayism so extreme that all the able political adventurers achieve their irresistible successes and overwhelming 97% plebiscite majorities by denouncing Liberty, Democracy, Opposition as putrefying anti-social superstitions.'[37] One of the early chapters of *Everybody's Political What's What?*, entitled 'The British Party System', gives the story of the origin of the party system in the form of 'a little historical drama', a playlet in two scenes with Sunderland, William III, Sunderland's son, and Walpole as the characters (and 25 years separating the scenes). Then Shaw comments:

[G]overnment by Parliaments modelled on the British Party System, far from being a guarantee of liberty and enlightened progress, must be ruthlessly discarded in the very fullest agreement with Oliver Cromwell, Charles Dickens, John Ruskin, Thomas Carlyle, Adolf Hitler, Pilsudski, Benito Mussolini, Stalin, and everyone else who has tried to govern efficiently and incorruptly by it, or who has studied its operation with a knowledge of its history and that of the Industrial Revolution. Contrast what it has done with what an efficient and entirely public spirited government might and should have done during the two centuries of its deplorable existence, or with what the Russian Soviet government has done in twenty years, and all our Whig Macaulayism drops dead before the facts.[38]

Shaw's distance from Macaulay is also evident in *The Thing Happens* in *Back to Methuselah*. When the fatuous prime minister, Burge-Lubin, tells the efficient, knowledgeable civil servant Confucius, 'You do not know the history of this country. . . . England once saved the liberties of the world by inventing

---

[36] Shaw, 'Something Like a History of England', *Pen Portraits and Reviews*, 93.
[37] Shaw to Winston Churchill, 8 May 1934, in Martin Gilbert, *Winston S. Churchill*, v, Companion Part 2, *The Wilderness Years 1929–1935* (London: Heinemann, 1981) 785.
[38] *Everybody's Political What's What?*, 23–9.

parliamentary government, which was her peculiar and supreme glory', Confucius replies, 'I know the history of your country perfectly well. It proves the exact contrary', and he demonstrates that the Macaulayite glories of England—parliamentary government, democracy, the Habeas Corpus Act, and trial by jury—are spurious (*CPP* v. 444–5).

Nor was Shaw impressed by what Macaulay regarded as the greatest achievement of the nineteenth century: the Reform Bill of 1832. For Macaulay this was the culmination of the movement towards parliamentary government that had its significant origins in the 1688 Revolution. For Shaw it 'inaugurated the purse-proud reign of the English middle class under Queen Victoria'.[39] For Macaulay the nineteenth century was the best of times. For Shaw, as for Carlyle, Ruskin, and Morris, it was the worst of times. Shaw described it as 'this wickedest of all the centuries', and referred to 'that darkest of dark ages, the nineteenth century'; and he wrote that G. K. Chesterton's *History of England* should have included an account of the establishment of Sir Robert Peel's police force: 'Without that new force the nineteenth century, rightly perceived by Mr Chesterton to have been the most villainous and tyrannous period in recorded history, could never have consummated its villainy in the full conviction that it was the proud climax of progress, liberty, and leaping and bounding prosperity. When its attention was drawn by some sensational horror to the cruellest and most bigoted of its own laws, it called them medieval, and believed it. What a theme for Mr Chesterton!'[40] If Shakespeare's history plays embody a Tudor myth that sanctions and glorifies the position of Elizabeth, then one could say that Shaw's work embodies an anti-Victorian myth, which calls into question nineteenth-century assumptions about the way in which all of history has led to the achievements of Victorian England.

In Shaw's view, the nineteenth century is not pre-eminently the age of parliamentary democracy, but rather the age of Capitalism. Shaw recognized that there were opponents of Capitalism in the Victorian age itself: 'Ruskin, Carlyle, and Dickens would

---

[39] *Intelligent Woman's Guide*, 215.
[40] *Shaw's Music*, i. 904; *Intelligent Woman's Guide*, 385; 'Something Like a History of England', *Pen Portraits and Reviews*, 93. There are many other formulations of this kind in Shaw's work.

have none of Macaulay's cheerful meliorism and progress-boosting: they saw that Capitalism was the robber's road to ruin'; and he said that attitudes towards Capitalism changed in the course of the century: '[I]t began to lose its moral plausibility, and, in spite of its dazzling mechanical triumphs and financial miracles, steadily progressed from inspiring the sanguine optimism of Macaulay and his contemporaries to provoking a sentiment which became more and more like abhorrence among the more thoughtful even of the capitalists themselves.'[41] Nevertheless, it was in the nineteenth century that Capitalism reached its most appalling excesses, and Shaw wrote that as soon as Marx made him realize this truth, his way of looking at the world changed. There is an unpublished letter that Shaw, in the last year of his life, sent to Gilbert Murray, which provides such a good summing up of this change in Shaw's view at the beginning of his adult life that it is worth quoting here at length:

> One of the differences between our retrospect of the nineteenth century is that you still accept the Whig view that it was a period of prosperity and happiness, hideously broken up by the wars of the twentieth. My outlook was much the same as yours until at the beginning of the 8oties Marx changed the mind of Europe and, for me, snatched the lid off hell and convinced me that the Manchester School, with its dreams and dogmas of Free Trade, Liberty, Equality and Fraternity, which satisfied Macaulay and Gladstone, blinded them to the horrors of the nineteenth century as the most damnably wicked and ruinous episode in human history. I had been quite pleased with Gladstone's Midlothian speeches, and admired those of John Bright. I was full of Mill's Essay on Liberty, Spencer's Data of Ethics, and Darwin. I was intensely anti-clerical; called myself publicly an Atheist at a meeting of the Shelley Society; and mistook that reaction for genuine Freethinking.
>
> The moment Marx got his knife into me, all this faded into an obsolete past; and I was thenceforth hated by the Liberals whom I discarded as fossils, and, with Sidney Webb, made the old Socialism, with its Liberal barricaders, constitutional by Fabianism, which is now, by the way, Stalinism in Russia.[42]

---

[41] *Everybody's Political What's What?*, 349; *Intelligent Woman's Guide*, 500.
[42] Shaw to Gilbert Murray, 5 Nov. 1949. MS in Bodleian Library, Oxford (Shaw–Murray Correspondence, fo. 175).

Here again we have Marx as the supplanter of Macaulay, and the Victorian era as a historical trough rather than a glorious point of culmination. The second half of the nineteenth century was also the era of Darwinism, which Shaw regarded as another sign of perdition. This Darwinist irreligion persisted into the twentieth century, and in the second half of Shaw's life he generally included the twentieth century with the nineteenth as the nadir of history. Thus in *The Thing Happens* in *Back to Methuselah* (1918–20), Burge-Lubin, in the year 2170, refers to the nineteenth and twentieth centuries as the Dark Ages (*CPP* v. 446); and the last part of the cycle, *As Far as Thought Can Reach*, reveals that the lowest point in human history was reached at the time of the First World War—this is when Lilith considered supplanting the human race with a new form of life (*CPP* v. 629). Writing just after the First World War, then, Shaw regarded this as the lowest point, but in 1948–9, when writing the Preface to *Farfetched Fables*, he had later horrors to take into account, and his judgement then was that there had certainly been no improvement in the last part of his life. 'The nineteenth century,' he said, 'which believed itself to be the climax of civilization, of Liberty, Equality, and Fraternity, was convicted by Karl Marx of being the worst and wickedest on record; and the twentieth, not yet half through, has been ravaged by two so-called world wars culminating in the atrocity of the atomic bomb' (*CPP* vii. 427–8). This does not sound very much like Macaulay's theory of progress.

4

Progress an Illusion?

> Enough, then, of this goose-cackle about Progress.
> (The Revolutionist's Handbook, *Man and Superman*,
> *CPP* ii. 774.)

> Are we agreed that Life is a force which has made innumerable experiments in organizing itself; that the mammoth and the man, the mouse and the megatherium, the flies and the fleas and the Fathers of the Church, are all more or less successful attempts to build up that raw force into higher and higher individuals, the ideal individual being omnipotent, omniscient, infallible, and withal completely, unilludedly self-conscious: in short, a god?
> (Don Juan in the Hell Scene of *Man and Superman*,
> *CPP* ii. 661–2.)

> The elements of progress and decline [are] strangely mingled in the modern mind.
> (Ruskin, *Modern Painters*.*)

The most notorious characteristic of Shaw's history plays is their liberal use of conspicuous and deliberate anachronisms. The dialogue that opens the original beginning of *Caesar and Cleopatra* ('An Alternative to the Prologue'), as the soldiers play at dice, forcibly unites past and present:

BELZANOR. By Apis, Persian, thy gods are good to thee.
THE PERSIAN. Try yet again, O captain. Double or quits!

(*CPP* ii. 169.)

Then in Act II, recent and ancient history are forced together when Caesar suggests Cyprus as a gift to Cleopatra's brother Ptolemy. Twenty years before the writing of the play Disraeli had acquired Cyprus for England at the end of the Russo-Turkish War, and here his celebrated phrase on his return from

---

* John Ruskin, *Modern Painters*, iii, *The Works of John Ruskin*, ed. E. T. Cook and Alexander Wedderburn, v (London: George Allen, 1904), 327 (Part IV, chap. XVI).

the Congress of Berlin in 1878 turns up in first century B.C. Alexandria:

POTHINUS [*impatiently*]. Cyprus is of no use to anybody.
CAESAR. No matter: you shall have it for the sake of peace.
BRITANNUS [*unconsciously anticipating a later statesman*]. Peace with honor, Pothinus.
(*CPP* ii. 203.)

This is followed immediately by an allusion to the English occupation of Egypt in the late nineteenth century, as some of Ptolemy's courtiers cry out 'Egypt for the Egyptians' (*CPP* ii. 204), which Shaw identified for his French translators as a phrase of the Liberal politician Sir William Harcourt.[1] Then later in the act the audience is reminded of London in Caesar's remark that the building next door, the theatre, 'commands the strand' (*CPP* ii. 217). Caesar's secretary, Britannus, is altogether the nineteenth-century Englishman rather than the ancient Briton, and Apollodorus, the aesthete, is more a figure of the 1890s than of the ancient world; his motto is 'Art for Art's sake' (Act III, *CPP* ii. 226). We are brought right up to date again in the fourth act when Ftatateeta chides Cleopatra, 'You want to be what these Romans call a New Woman' (*CPP* ii. 255). There is a nice Shavian touch in Act V: Apollodorus says that Caesar 'was settling the Jewish question when I left' (*CPP* ii. 285), which sounds like an anachronistic intrusion but turns out to be taken straight from Mommsen, who said that after his military victory in Alexandria, Caesar 'contented himself with granting to the Jews settled in Alexandria the same rights which the Greek population of the city enjoyed'.[2]

As Martin Meisel has pointed out, Shaw's anachronisms in *Caesar and Cleopatra* are in the tradition of nineteenth-century burlesque and extravaganza (Offenbach's *opéra bouffe*, *La Belle Hélène*, for example),[3] but the anachronisms serve many other

---

[1] Shaw's MS revisions in the typescript of A. and H. Hamon's French translation of *Caesar and Cleopatra*, Harry Ransom Humanities Research Center, the University of Texas at Austin.
[2] Theodor Mommsen, *The History of Rome*, iv (tr. William P. Dickson; New York: Charles Scribner's Sons, 1891), 516.
[3] Martin Meisel, *Shaw and the Nineteenth-Century Theater* (Princeton: Princeton Univ. Press, 1963), 386.

purposes as well as introducing an element of the ludicrous into Shaw's plays. In his 1944 Postscript to *Back to Methuselah*, Shaw paraphrased the opening of St John's Gospel ('In the beginning was the Word . . .') as 'Attention is the first symptom of thought', and he added: 'John the Evangelist would have worded it so, had he been born a Victorian' (*CPP* v. 699). Although Shaw sees sharp distinctions between one epoch and another, there is also a continuity in human experience that creates links between different historical periods, and thus we have the assumption here that in certain cases it may be the language which changes rather than the ideas. It is the same assumption, which lies behind many of the anachronisms in Shaw's plays, that is brought out in a joke in the first act of *Major Barbara*. Lady Britomart asks Cusins to translate Lomax's inane remark, 'Well, you must admit that this is a bit thick', into 'reputable English', and Cusins obliges: 'I think Charles has rather happily expressed what we all feel. Homer, speaking of Autolycus, uses the same phrase. πυκινὸν δόμον ἐλθεῖν means a bit thick' (*CPP* iii. 81). Shaw's taste for anachronistic slang goes right back to his first attempt at a play, the *Passion Play* that he wrote in 1878, when Judas appears at Joseph's workshop, wanting some repairs done, and Jesus advises him to 'take it elsewhere. We would but botch it' (Act I, *CPP* vii. 494). Rufio's line in Act II of *Caesar and Cleopatra*, '[Y]ou may save your breath to cool your porridge' (*CPP* ii. 210), is repeated 25 years later by the Dauphin in *Saint Joan* (Scene ii, *CPP* vi. 112), which of course is filled with linguistic anachronisms of this type, particularly in the speeches of the title character.

One of the functions of slang and other kinds of anachronism is to reveal the kinship between past and present, and thus to give the past a reality for present-day audiences. 'History', wrote G. M. Trevelyan, 'starts out from this astonishing proposition —that there is no difference in degree of reality between past and present.'[4] This is precisely the proposition that Shaw's history plays start out from as well. Shaw wanted his history plays to possess the quality that Carlyle attributed to the novels of Walter Scott. Scott's historical novels, according to Carlyle, 'have taught all men this truth, which looks like a truism, and yet was as good

---

[4] George Macaulay Trevelyan, *Clio, A Muse and Other Essays* (London: Longmans, Green, 1930), 100.

as unknown to writers of history and others, till so taught: that the bygone ages of the world were actually filled by living men, not by protocols, state-papers, controversies and abstractions of men. Not abstractions were they, not diagrams and theorems; but men, in buff or other coats and breeches, with colour in their cheeks, with passions in their stomach, and the idioms, features and vitalities of very men.'[5] Scott's ways of conveying the reality of historical figures are very different from Shaw's, but the anachronisms in Shaw's plays serve something of the same purpose as the mass of detail or the attention to speech patterns in Scott's novels. A nineteenth-century historical writer whose methods are akin to those of Shaw is Theodor Mommsen. When his *History of Rome* was accused of lacking tranquillity and dignity, he defended his idiomatic usages in terms that anticipate Shaw's position: 'Much might be said about the modern tone. I wanted to bring down the ancients from their fantastic pedestal into the real world. That is why the consul had to become the burgomaster.'[6]

Such was Shaw's attitude towards the popular perception of Jesus. In the Preface to *Androcles and the Lion* he discussed Jesus and his ideas in terms of the present, and he argued that it is salutary for people to have their god brought into the real world. His comments on 'the idolatrous or iconographic worship of Christ' may still retain some power to shock:

> By this I mean literally that worship which is given to pictures and statues of him, and to finished and unalterable stories about him. The test of the prevalence of this is that if you speak or write of Jesus as a real live person, or even as a still active God, such worshippers are more horrified than Don Juan was when the statue stepped from its pedestal and came to supper with him. You may deny the divinity of Jesus; you may doubt whether he ever existed; you may reject Christianity for Judaism, Mahometanism, Shintoism, or Fire Worship; and the iconolaters, placidly contemptuous, will only classify you as a freethinker or a heathen. But if you venture to wonder how Christ would have looked if he had shaved and had his hair cut, or what size in shoes he took, or whether he swore when he stood on a nail in the carpenter's shop, or could not button his robe when he was in a hurry, or whether he laughed

[5] Thomas Carlyle, 'Sir Walter Scott', *Critical and Miscellaneous Essays*, iv. 77–8, *The Works of Thomas Carlyle*, ed. H. D. Traill (Centenary Edn., 30 vols., London: Chapman and Hall, 1896–9), xxix.

[6] Quoted in G. P. Gooch, *History and Historians in the Nineteenth Century* (1913; London: Longmans, 1952), 461.

over the repartees by which he baffled the priests when they tried to trap him into sedition and blasphemy, or even if you tell any part of his story in the vivid terms of modern colloquial slang, you will produce an extraordinary dismay and horror among the iconolaters. You will have made the picture come out of its frame, the statue descend from its pedestal, the story become real.   (*CPP* iv. 514–15.)

Similarly in *Androcles and the Lion* itself the conflict between the Christians and Romans is brought home to us as a familiar matter that transcends time, and the play tramples on the pious paintings of martyrdom that some people would associate with its subject. In a letter to a newspaper at the time of the play's first production in 1913, Shaw said that most people 'are shocked at the idea of [the Christians] being callously called to their deaths as numbered turns in a variety entertainment by a vulgar call-boy, instead of simply being painted by Royal Academicians as being politely led up to heaven by angels with palm branches' ('Androcles: How Divines Differ about Shaw', *CPP* iv. 648). Shaw's martyrdom play pulls the picture out of its frame, topples the statue down from its pedestal, and makes the story become real.

One might say that Shaw, like Mommsen, wanted to write history in the present tense. He wanted to make his audiences see that historical figures are real people, and that historical issues are real issues. He said that it gave him an 'extraordinary satisfaction' when the outrage of conventional audiences at *Androcles and the Lion* proved 'that I have brought them face to face for the first time with the grim reality of persecution and their own daily complicity in it' ('Androcles: How Divines Differ about Shaw', *CPP* iv. 648). To write about the past as if it were the present makes an audience aware of the reality of the past, and it also gives them a better understanding of contemporary issues. This attitude of Shaw's would not have earned the approval of Herbert Butterfield, who said in *The Whig Interpretation of History* that 'The study of the past with one eye, so to speak, upon the present is the source of all sins and sophistries in history, starting with the simplest of them, the anachronism.'[7] Leaving aside the value

[7] H. Butterfield, *The Whig Interpretation of History* (1931; New York: Norton, 1965), 31–2. The *English Historical Review*, at its inception in 1886, announced a policy of 'refusing contributions which argue . . . questions with reference to present controversy' (quoted in Hayden White, *Metahistory: The Historical Imagination in Nineteenth-Century Europe* (Baltimore and London: The Johns Hopkins Univ. Press, 1973), 137).

judgement here, we have a good description of Shaw's approach to history.

Shaw's sense of the relevance of the past is like Carlyle's. To Shaw the past is not something separate from the present, but rather a way of looking at the present. The past has the same degree of reality as the present, and also the same degree of relevance to contemporary human experience. He would have agreed with Carlyle's opinion that the seventeenth century is worthless except in so far as it can be made the nineteenth (or twentieth), and his practice as a historian is similar to Carlyle's in *Past and Present*, using the past as a way of drawing people's attention to the problems and the underlying realities of the present. At the end of the first chapter of Carlyle's book, as he is about to move from industrial England in the 1840s to feudal England in the twelfth century, he explains his method. As 'editor' of the documents revealing the reality of the medieval monastery, he

> will endeavour to select a thing or two; and from the Past, in a circuitous way, illustrate the Present and the Future. The Past is a dim indubitable fact; the Future too is one, only dimmer; nay properly it is the *same* fact in new dress and development. For the Present holds it in [sic] both the whole Past and the whole Future;—as the LIFE-TREE IGDRASIL, wide-waving, many-toned, has its roots down deep in the Death-kingdoms, among the oldest dead dust of men, and with its boughs reaches always beyond the stars; and in all times and places is one and the same Life-tree![8]

Carlyle, like Shaw, is able to use the past in the way he does because he accepts no essential distinction between past and present. All moments in time (including the future as well) are part of one continuum.

This conception of organic continuity in history underlies a passage like the 1912 Prologue to *Caesar and Cleopatra*, in which Ra ensures that no member of the audience will respond to ancient history as something unconnected with his own society. Ra, speaking across the centuries directly to his modern English audience, makes the connection between Roman imperialism in Caesar's day, and British imperialism in the present. It is revealing to see that when Shaw adapted this Prologue for a Polish

---

[8] Carlyle, *Past and Present*, *Works*, x. 38 (Book i, chap. vi).

audience, he changed the English references so that the audience would apply Ra's pronouncements to their own nation. Thus Ra talks about 'an old Poland and a new', and about 'this little Europe from the Oder to the heart of Muscovy', and a reference to the Elizabethans' victory over Spain is changed to the crumbling of the power of Islam 'when it was set against your fathers in the days when Poland knew her mind'.[9]

Whereas *Caesar and Cleopatra* begins with a reminder of the contemporary relevance of its subject, *Androcles and the Lion* concludes with an essay which has the same intention. Writing during the First World War, Shaw noted a report that the German Crown Prince had walked out of a performance of the play when it was first done in Berlin: 'No English Imperialist was intelligent and earnest enough to do the same in London. If the report is correct, I confirm the logic of the Crown Prince, and am glad to find myself so well understood. But I can assure him that the Empire which served for my model when I wrote Androcles was, as he is now finding to his cost, much nearer my home than the German one' (*CPP* iv. 641). In this expository conclusion to *Androcles and the Lion* Shaw drew analogies between ancient Roman and modern British persecution, and in a note written for the New York production of the play in 1915 he made the point that 'None of the characters are monsters: they are just such people as may be found in the United States today, placed in the monstrous circumstances created by the Roman Empire' ('A Note to "Androcles and the Lion"', *CPP* iv. 651).

There is much of this sort of argument in the Preface to Shaw's later play about religious persecution, *Saint Joan*, where the author explained the nature of his interest in Joan's history. The successor to the medieval Church is the modern medical profession, which is much more tyrannical, and 'Therefore the question raised by Joan's burning is a burning question still, though the penalties involved are not so sensational. That is why I am probing it. If it were only an historical curiosity I would not waste my readers' time and my own on it for five minutes' (*CPP* vi. 58). In a BBC talk about Joan of Arc in 1931 (on the five hundredth anniversary of her death), Shaw offered Leon Trotsky as a

---

[9] Shaw, 'Prologue for Poland', MS in Harry Ransom Humanities Research Center, the University of Texas at Austin.

contemporary analogy to Joan, and he asserted that 'the whole value of Joan to us is how you can bring her and her circumstances into contact with our life and our circumstances' ('Saint Joan: A Radio Talk', *CPP* vi. 226–7, 229).

The principal function of the anachronisms in Shaw's plays, I think, is to demonstrate how much of history remains unchanged. In addition to their pedagogical function of making an audience see the reality and relevance of the past, they convey Shaw's resistance to the theory of progress that dominated late Victorian attitudes towards history. Epochs are distinct, but nothing *essential* has altered in that human beings have not improved morally or intellectually. Shaw draws past and present together in order to convey to his audience the Enlightenment view of an unchanging human nature—or, to be more precise, the Shavian view of a hitherto unchanging human nature. In some respects Shaw's history plays point to the truth of Voltaire's judgement that 'man, generally speaking, was always what he is now', and also Hume's opinion on this question: 'It is universally acknowledged that there is a great uniformity among the actions of men, in all nations and ages, and that human nature remains still the same, in its principles and operations. . . . Would you know the sentiments, inclinations, and course of life of the Greeks and Romans? Study well the temper and *actions* of the French and English.'[10] Shaw's history plays convey something of the distinctiveness of different epochs, and at the same time they invite their audiences to study the temper and actions of the Romans, Egyptians, and others in order to know the sentiments of the contemporary French and English. The plays reveal as illusory the audience's notion that they are more advanced than previous generations.

This idea that there has been no improvement over the centuries is also conveyed by a pervasive rhetorical strategy in Shaw's expository writing: the historical analogy, in which the prose will leap over historical time in a way that implies there has been no significant change from one era to another. An argument that the most virtuous people are not necessarily the best rulers provides

---

[10] Voltaire, *Essai sur les moeurs*, quoted in Isaiah Berlin, *Vico and Herder* (London: Hogarth Press, 1976), 197; David Hume, *Enquiry Concerning Human Understanding*, quoted in Louis O. Mink, 'Narrative Form as a Cognitive Instrument', in Robert H. Canary and Henry Kozicki, eds., *The Writing of History* (Madison: Univ. of Wisconsin Press, 1978), 138.

Progress an Illusion?

one good example of the Shavian historical sweep: 'Aknaton, in the 14th century before Christ, and Amanullah yesterday, came to grief as monarchs as hopelessly as our pious simpleton James II.'[11] This sentence, which de-emphasizes chronology in jumping from the beginning of history to the Afghan king who was deposed in 1929 and from there to English history of the seventeenth century, illustrates a very common form of exposition in Shaw's work. Often this type of construction will use historical figures and events in order to explain the contemporary world, as in a discussion of Soviet Russia: 'Lenin and his successors were not able to extricate the new Russian national State they had set up from this new Russian international (Catholic) Church any more than our Henry II or the Emperor who had to come to Canossa was able to extricate the English State and the medieval Empire from the Church of Rome.'[12] The 'any more than' indicates a historical equivalence, as does the 'exactly as' in the following: '. . . [T]he Russian Government imprisons Bourtseff, and arrests Socialist members of the Duma exactly as Charles I tried to arrest members of the British House of Commons two and a half centuries ago.'[13]

The third chapter of Macaulay's *History of England*, which gives a picture of English life at the time of Charles II's death, juxtaposes past and present in order to point up the differences between 1685 and 1848. The impression we are given is one of ascent. Shaw's juxtapositions leave a very different impression. They point up the similarities between past and present, and their effect is one of historical levelling. One passage that makes this

---

[11] Shaw, *Everybody's Political What's What?* (London: Constable, 1944), 323.

[12] Shaw, *The Intelligent Woman's Guide to Socialism, Capitalism, Sovietism and Fascism* (London: Constable, 1949), 442.

[13] Shaw to Maxim Gorki, 28 Dec. 1915, *Collected Letters 1911–1925*, ed. Dan H. Laurence (London: Max Reinhardt, 1985), 342. Cf. 'Henry VIII, a royal Leader, plundered the Church and made it a crime to be a Catholic priest; but he immediately had to disgorge his booty and distribute it among his prefects and their families. *In precisely the same way* Führer Hitler has plundered the Jews and made it a crime to be a Jew in Germany. But he, too, has had to leave their jobs and their belongings to be owned and exploited by German employers who are sweating the German proletariat as rapaciously as any Jew' (*Intelligent Woman's Guide*, 485–6, italics added). For a discussion of the ways in which '[h]istorical cross-reference and other patterns of Shavian comparison are sharply etched in [Shaw's] style', see Richard M. Ohmann, *Shaw: The Style and the Man* (Middletown, Conn.: Wesleyan Univ. Press, 1962), 13–21.

levelling explicit is a discussion in *The Intelligent Woman's Guide to Socialism* of the possible future domination of Parliament by the Labour Party. 'The danger is that it may split into half a dozen or more irreconcilable groups, making parliamentary government impossible', Shaw warned. 'That is what happened in the Long Parliament in the seventeenth century, when men were just what they are now, except that they had no telephones nor airplanes. The Long Parliament was united at first by its opposition to the King. But when it cut off the King's head, it immediately became so disunited that Cromwell, like Signor Mussolini today, had at last to suppress its dissensions by military force, and rule more despotically than ever the king had dared.'[14] To Macaulay the telephones and aeroplanes would make a difference, and improved political wisdom and constitutional stability would make a repetition of the events of the 1650s out of the question.

Shaw, on the other hand, is always eager to show that, because there has been no improvement, history repeats itself. In his discussions of toleration and persecution, for example, he demonstrates that (contrary to the Whig reading of history) we are no better than our ancestors. '[M]y martyrs are the martyrs of all time, and my persecutors the persecutors of all time,' he said in his note at the end of *Androcles and the Lion*, and this note makes it clear that 'all time' includes the present time. In an area in which most people in 1916 (when the play was first published in England) would have thought there had been enormous improvement since Roman times, Shaw insisted that there had been no improvement whatsoever. '[I]f the Government decided to throw persons of unpopular or eccentric views to the lions in the Albert Hall or the Earl's Court stadium tomorrow, can you doubt that all the seats would be crammed,' he asked. The twentieth-century martyrs would not be religious heretics, but rather 'Peculiars, Anti-Vivisectionists, Flat-Earth men, scoffers at the laboratories, or infidels who refuse to kneel down when a procession of doctors goes by' (*CPP* iv. 636, 640). This parallel between religious persecution of the past and scientific or medical tyranny of the present is one of the recurring themes in Shaw's expository prose. 'Dread of epidemics: that is, of disease and premature death, has

[14] *Intelligent Woman's Guide*, 344–5.

created a pseudo-scientific tyranny just as the dread of hell created a priestly tyranny in the ages of faith', he wrote,[15] linking together present and past defects with the characteristic 'just as'. This argument that the present is no better than the past with respect to persecution also runs through the Preface to *Saint Joan*. One of the many assertions along these lines is that Socrates' accuser, 'if born 2300 years later, might have been picked out of any first class carriage on a suburban railway during the evening or morning rush from or to the City', while another passage asserts 'that there was not the smallest ground for the self-complacent conviction of the nineteenth century that it was more tolerant than the fifteenth, or that such an event as the execution of Joan could not possibly occur in what we call our own more enlightened times' (*CPP* vi. 16, 61).

The chief exponent of this 'self-complacent conviction of the nineteenth century', Macaulay, has a digression in his essay on Lord Bacon in which he argues that, contrary to common belief, women of the present day are better educated than women of the sixteenth century. 'When ... we compare the acquirements of Lady Jane Grey with those of an accomplished young woman of our own time, we have no hesitation in awarding the superiority to the latter.' He wants to convince his readers that 'they are mistaken in thinking that the great-great-grandmothers of their great-great-grandmothers were superior women to their sisters and their wives'.[16] Shaw's object is to convince his readers that they are mistaken in thinking themselves superior to their ancestors: 'Do not be deceived by modern professions of toleration. Women are still what they were when the Tudor sisters sent Protestants to the stake and Jesuits to the rack and gallows; when the defenders of property and slavery in Rome set up crosses along the public roads with the crucified followers of the revolted gladiator slave Spartacus dying horribly upon them in thousands; and when the saintly Torquemada burnt alive every Jew he could lay hands on as piously as he told his beads.'[17]

Against Macaulay's view that the past was no better than the

---

[15] *Intelligent Woman's Guide*, 398.
[16] T. B. Macaulay, 'Lord Bacon', *The Works of Lord Macaulay*, ed. Lady Trevelyan (8 vols., London: Longmans, Green, 1879), vi. 146.
[17] *Intelligent Woman's Guide*, 368–9.

present, then, we have Shaw's view that the present is no better than the past, and whereas Macaulay goes a step further and exalts the present over the past, Shaw goes a step further in *his* direction and exalts the past over the present. That is, Shaw rejects the Macaulayite theory of ascent to the extent of countering it with a theory of decline. His usual view is the levelling one that we have been looking at, but when he does allow for a significant change between past and present it is a change for the worse. Thus the Preface to *Saint Joan* not only argues that we are just as bad as the Middle Ages and other past eras when it comes to persecution, but it sometimes puts this more emphatically by arguing that we are worse. Today 'legal compulsion to take the doctor's prescription, however poisonous, is carried to an extent that would have horrified the Inquisition and staggered Archbishop Laud', and 'Joan got a far fairer trial from the Church and the Inquisition than any prisoner of her type and in her situation gets nowadays in any official secular court' (*CPP* vi. 58, 19).

In a review of *Aïda* at Covent Garden in 1888 Shaw complained that the conductor, Signor Mancinelli, 'conducted the court and temple scenes barbarically, evidently believing that the ancient Egyptians were a tribe of savages, instead of, as far as one can ascertain, considerably more advanced than the society now nightly contemplating in "indispensable evening dress" the back of Signor Mancinelli's head'.[18] Ten years later he began *Caesar and Cleopatra* with a similar assertion about the relationship between past and present, in the stage direction that opens the 'Alternative to the Prologue': The palace in ancient Egypt *'is not so ugly as Buckingham Palace; and the officers in the courtyard are more highly civilized than modern English officers: for example, they do not dig up the corpses of their dead enemies and mutilate them, as we dug up Cromwell and the Mahdi'* (*CPP* ii. 168).[19] There is a comparable passage in the opening stage

[18] *Shaw's Music*, ed. Dan H. Laurence (3 vols., London: Max Reinhardt, The Bodley Head, 1981), i. 519.

[19] At the time Shaw was working on *Caesar and Cleopatra* he wrote a letter to a newspaper about Lord Kitchener's army's destruction of the tomb of the Mahdi (the Sudanese leader, who had died in 1885). Shaw said that when he first heard the story, he assumed that it was merely propaganda, a re-working of the 'history of the body of Oliver Cromwell, which, at the Restoration, was dug up from its grave in Westminster Abbey, mutilated, and exposed at Tyburn and Temple Bar. This piece of history has

direction in Act II, describing the royal palace at Alexandria: an English manufacturer would not appreciate the place because 'Tottenham Court Road civilization is to this Egyptian civilization as glass bead and tattoo civilization is to Tottenham Court Road' (*CPP* ii. 195).

The Prologue that Shaw provided in 1912 to replace the original opening of the play is one of the major statements of his historical attitudes, and it too elevates the past above the present—or at least subordinates the present to the past. Ra proclaims to his modern audience that as the ancient Egyptians were 'so ye are; and yet not so great; for the pyramids my people built stand to this day; whilst the dustheaps on which ye slave, and which ye call empires, scatter in the wind even as ye pile your dead sons' bodies on them to make yet more dust' (*CPP* ii. 162). Modern England is inferior to ancient Egypt, and also to Elizabethan England, for Ra refers to the defeat of Spain 'in the days when England was little, and knew her own mind, and had a mind to know instead of a circulation of newspapers' (*CPP* ii. 164). The message of both Ra's Prologue and the Alternative to the Prologue is that nothing has changed very much in the past two thousand years, and in so far as there have been changes they have been for the worse. In *Caesar and Cleopatra* it is important that an audience not feel superior to the Egyptians, for we are to feel the Egyptians' inferiority to Caesar as our own. Caesar towers above the other Egyptians and Romans in the play, just as he would tower above us today. One of the strategies of the play is to prevent us from patronizing him as a man of the past. The play therefore makes it clear that we are at the level of Cleopatra and the others, and that from the time of Caesar to the present there has been no progress, and if anything there has been decline.

Another view of Shaw's that runs counter to the Victorian theory of progress is his belief in the recurring collapse of civilizations. History does not consist of a progressive movement

---

hitherto been used to mark the Tartar-like savagery of English society 200 years ago. Macaulay's comment on it is well known' (*Agitations: Letters to the Press 1875–1950*, ed. Dan H. Laurence and James Rambeau (New York: Frederick Ungar, 1985), 49–50). Macaulay's comment is in chapter 11 of *The History of England*: 'Cromwell was no more; and those who had fled before him were forced to content themselves with the miserable satisfaction of digging up, hanging, quartering, and burning the remains of the greatest prince that has ever ruled England' (*Works*, i. 122).

Progress an Illusion?

from one civilization to another, but rather of a series of civilizations that reach a certain point and then disintegrate. In his lecture in New York City in 1933 he explained the change that had come about in our historical perspective since the Victorian era:

> Within my lifetime our knowledge of history has been greatly extended. We used to be taught that antiquity meant the Roman Empire, which had absorbed the Greek city States with the pyramids of Egypt looking on, and with Jerusalem and a sketchy Babylonian collection of idolators in the hinterland. The one belief that we got out of it all was that modern civilization was an immense improvement on those barbarous times, and that all white people had been steadily progressing, getting less and less superstitious, less and less savage, more and more enlightened, until the pinnacle had been reached, represented by ourselves.
>
> We are now beginning to have serious doubts whether we ourselves are in any way remarkable or unprecedented as specimens of political enlightenment; for our new knowledge of history tells us that our picture of the past was false. Thanks largely to the researches of Professor Flinders Petrie, we know of five or six ancient civilizations which were just like our own civilization, having progressed in the same way, to the same artistic climaxes, the same capitalistic climaxes, the same democratic and feminist climaxes as we; and they all perished. . . .
>
> That puts us in a very different mental attitude from our fathers and grandfathers, because what we are up against now is the fact that we too have reached the edge of the precipice over which these civilizations fell and were dashed to pieces.[20]

Flinders Petrie, Professor of Egyptology at University College London, is mentioned several times in Shaw's work in this context of collapsing civilizations, and Shaw's moral is always the same: that we will be next unless we radically change the nature of our society and ourselves. In *The Gospel of the Brothers Barnabas* in *Back to Methuselah*, for example, the biologist Conrad Barnabas tells the politicians that our present lifespan is inadequate. 'Flinders Petrie has counted nine attempts at civilization made by people exactly like us; and every one of them failed just as ours is failing. They failed because the citizens and statesmen died of old age or over-eating before they had grown

---

[20] Shaw, *The Political Madhouse in America and Nearer Home* (London: Constable, 1933), 27–8.

out of schoolboy games and savage sports and cigars and champagne' (*CPP* v. 419–20).

Here the cause of collapse is an insufficient lifespan, but actually in different works Shaw expresses the cause differently according to the needs of particular arguments. In 1889, in his account of 'The Economic Basis of Socialism', it was private property: 'All attempts yet made to construct true societies upon it have failed: the nearest things to societies so achieved have been civilizations, which have rotted into centres of vice and luxury, and eventually been swept away by uncivilized races.'[21] In the Epistle Dedicatory to *Man and Superman*, published in 1903, it was the need for a better breed of citizen: '[W]e must get an electorate of capable critics or collapse as Rome and Egypt collapsed. At this moment the Roman decadent phase of *panem et circenses* is being inaugurated under our eyes' (*CPP* ii. 515); and in the Revolutionist's Handbook the incapacity of modern governments will lead to 'Ruins of Empires, New Zealanders sitting on a broken arch of London Bridge [as in Macaulay's essay on Ranke], and so forth' (*CPP* ii. 755). In 'The Common Sense of Municipal Trading' (1904), it was 'the phenomenon of economic rent, that rock on which all civilizations ultimately split and founder'.[22] In the Preface to *Major Barbara* (1906), it was the toleration of poverty 'which has already destroyed so many civilizations, and is visibly destroying ours in the same way (*CPP* iii. 26), and in the Preface to *Misalliance* (written in 1914), it was '[a]ll this inculcated adult docility, which wrecks every civilization as it is wrecking ours' (*CPP* iv. 87). In *The Intelligent Woman's Guide to Socialism*, it was economic causes again: politicians and voters have begun 'clamoring that the existing distribution of wealth is so anomalous, monstrous, ridiculous, and unbearably mischievous, that it must be radically changed if civilization is to be saved from the wreck to which all the older civilizations we know of were brought by this very evil'.[23]

Another reader of Flinders Petrie was William Butler Yeats. In *A Vision* (1925, 1937), he listed Petrie's *The Revolutions of*

---

[21] Shaw, 'The Economic Basis of Socialism', *Essays in Fabian Socialism* (London: Constable, 1949), 23.
[22] Shaw, 'The Common Sense of Municipal Trading', *Essays in Fabian Socialism*, 217.
[23] *Intelligent Woman's Guide*, 5; see also 59, 96.

Civilisation as one of the sources for his historical system, and he mentioned him in a section dealing with the Great Wheel of 2,000-year cycles.[24] Shaw's habit of sweeping over history and levelling people and events of different periods, and also his view of the collapse of previous civilizations, would suggest a cyclical view of history, in which the dominant pattern is one of recurrence rather than the linear forward movement of progress. In the Preface to *The Dark Lady of the Sonnets*, however, he introduced Thomas Tyler, a former acquaintance who had published an edition of Shakespeare's sonnets, as 'a specialist in pessimism [who] ... delighted in a hideous conception which he called the theory of the cycles, according to which the history of mankind and the universe keeps eternally repeating itself without the slightest variation throughout all eternity' (*CPP* iv. 272). This description of Tyler's views certainly does not imply much sympathy on Shaw's part, and in the Hell Scene of *Man and Superman* the exponent of the theory of the cycles is the Devil himself. This theory forms the culmination of the Devil's argument against Juan's attempts to articulate a theory of progress. '[A]ll history', he declares, 'is nothing but a record of the oscillations of the world' between the extremes of heaven and hell.

THE DEVIL. ... An epoch is but a swing of the pendulum; and each generation thinks the world is progressing because it is always moving. But when you are as old as I am; when you have a thousand times wearied of heaven, like myself and the Commander, and a thousand times wearied of hell, as you are wearied now, you will no longer imagine that every swing from heaven to hell is an emancipation, every swing from hell to heaven an evolution. Where you now see reform, progress, fulfilment of upward tendency, continual ascent by Man on the stepping stones of his dead selves to higher things, you will see nothing but an infinite comedy of illusion.

(*CPP* ii. 683.)

Juan is not able to reject the Devil's cyclical theory of history. He grants that the Life Force 'has hit on the device of the clockmaker's pendulum, and uses the earth for its bob; that the history of each oscillation, which seems so novel to us the actors, is but the history of the last oscillation repeated' (*CPP* ii. 684). His question, 'has the colossal mechanism no purpose?', is one that I

---

[24] W. B. Yeats, *A Vision* (London: Macmillan, 1962), 261, 203 n.

want to return to shortly, but at this stage we can see that *Man and Superman* accepts a theory of historical cycles, and we can see this against the background of Shaw's various comments about repetitive patterns in history.

This idea of oscillation as opposed to progress is also conspicuous in Tanner's Revolutionist's Handbook, and the repetitive nature of things is implied in the treatment of the Don Juan legend in *Man and Superman*: figures from the past like Don Juan and Doña Ana live again as Tanner, Ann, and the others in turn-of-the-century London. The play suggests that relations between Man and Woman have not changed so much since the time of the Don Juan legend, and that Tanner and Ann are Juan and Ana in modern dress. In much the same way, Savvy in *The Gospel of the Brothers Barnabas* believes 'the old people are the new people reincarnated. . . . I suspect I am Eve. I am very fond of apples; and they always disagree with me', to which her uncle Conrad replies, 'You *are* Eve, in a sense. The Eternal Life persists; only It wears out Its bodies and minds and gets new ones, like new clothes. You are only a new hat and frock on Eve' (*CPP* v. 423).[25] The historical pattern in *Back to Methuselah* as a whole is one of recurrence. Adam and Eve bring death into the world, until the second Adam and Eve, the unlikely pair Haslam and the Parlor Maid, begin to reverse the process. The apparently upward movement within the five-play cycle is in part a movement towards the recovery of what was lost in the first play. In the final play, *As Far as Thought Can Reach*, people live until they have had their accident, as was the case in the first act of *In the Beginning*. As in *Man and Superman*, characters are reincarnated centuries later. Cain becomes General Aufsteig in AD 3000; Burge and Lubin, the politicians of the present day, merge into Burge-Lubin, President of the British Islands in the year 2170, who is reincarnated as Ambrose Badger Bluebin, Prime Minister of the British Islands in the year 3000. The pattern of recurrence is also reinforced by the reappearance, at the end of the last play, of the ghosts of all the characters from the first play. In the Preface to *Back to Methuselah*, there is a suggestion of the idea of cyclical recurrence in the assertion that 'History records very little in the way of mental activity on the part of the mass of mankind except

---

[25] Cf. Carlyle's imagery of clothes—e.g. the passage about the future as the past in new dress (quoted above, p. 106).

a series of stampedes from affirmative errors into negative ones and back again' (*CPP* v. 327).

'[W]hat the world calls originality is only an unaccustomed method of tickling it', Shaw wrote in the Preface to *Three Plays for Puritans* in order to explain how he achieved his reputation for originality by reviving discarded dramatic styles (*CPP* ii. 47), and he suggested the cyclical nature of aesthetic taste in a stage direction in one of the plays in this volume. The Andersons' sitting-room, in the eighteenth-century New Hampshire of *The Devil's Disciple*, is described at the beginning of the second act, and to the description this comment is added: '*On the whole, it is rather the sort of room that the nineteenth century has ended in struggling to get back to under the leadership of Mr Philip Webb* [William Morris's associate] *and his disciples in domestic architecture, though no genteel clergyman would have tolerated it fifty years ago*' (*CPP* ii. 84). Clearly, this is a kind of irony that Shaw relished, and it is an important element in his conception of historical development. It is really the same irony, in a different mode, that pervades *Androcles and the Lion*, where we are conscious that emergent Christianity, which was regarded as a revolutionary heresy under the Roman Empire, subsequently became the established religion of Europe, and thus equivalent to the official Roman religion in the play, and guilty of exactly the same kind of persecution of new heresies. The names change, but the fundamental patterns in human experience recur. New movements harden into shells, which must then be destroyed, at which point the cycle begins to repeat itself. This is change, but it is not progress.

One of the Maxims for Revolutionists appended to *Man and Superman* is that 'Those who admire modern civilization usually identify it with the steam engine and the electric telegraph' (*CPP* ii. 794). Mechanical inventions do not, in Shaw's judgement, constitute genuine progress. What Shaw looks for, and fails to find in history, is evidence of an improvement in the quality of human beings, in their intellectual and moral understanding. Examples of apparent progress, such as the Factory Acts, turn out to be 'merely the changes that money makes'. 'Still, they produce an illusion of bustling progress; and the reading class infers from them that the abuses of the early Victorian period no longer exist except as amusing pages in the novels of Dickens. But the moment

Progress an Illusion? 119

we look for a reform due to character and not to money, to statesmanship and not to interest or mutiny, we are disillusioned' (Revolutionist's Handbook, *Man and Superman, CPP* ii. 765). Andrew Undershaft in *Major Barbara* complains that the world is ready to scrap its obsolete steam engines and dynamos, 'but it wont scrap its old prejudices and its old moralities and its old religions and its old political constitutions. Whats the result? In machinery it does very well; but in morals and religion and politics it is working at a loss that brings it nearer bankruptcy every year' (Act III, *CPP* iii. 171). It is only changes in human character that would create the possibility of progress in morality, religion, and politics. One figure in Shaw's plays who believes in progress is the distasteful Sir George Crofts in *Mrs Warren's Profession*, who boasts to Vivie that instead of an adherence to any particular religion he has 'an honest belief that things are making for good on the whole' (*CPP* i. 326). He himself, as a thoroughly representative member of a corrupt society, is hardly evidence of moral, religious, or political improvement.

The distinction between mere mechanical improvement and real progress is made in the long note appended to *Caesar and Cleopatra* on 'Apparent Anachronisms', which is one of Shaw's main expressions of his anti-progressive bent. He said here that his reason for ignoring the popular conception of progress in the play was that there is no reason to suppose that any progress has taken place since the time of Caesar and Cleopatra. By the popular conception of progress he did not mean 'increased command over Nature' but rather an improvement in human beings, and 'the period of time covered by history is far too short to allow of any perceptible progress in the popular sense of Evolution of the Human Species. The notion that there has been any such Progress since Caesar's time (less than 20 centuries) is too absurd for discussion.' The common belief that there has been progress is a result of 'the ordinary citizen's ignorance of the past combine[d] with his idealization of the present'. The first paragraph of this 'Apparent Anachronisms' note offers a forceful rejection of the Macaulayite faith in progress and the Whig view of history:

The more ignorant men are, the more convinced are they that their little parish and their little chapel is an apex to which civilization and philosophy has painfully struggled up the pyramid of time from a desert of savagery. Savagery, they think, became barbarism; barbarism became ancient civilization; ancient civilization became Pauline Christianity; Pauline Christianity became Roman Catholicism; Roman Catholicism became the Dark Ages; and the Dark Ages were finally enlightened by the Protestant instincts of the English race. The whole process is summed up as Progress with a capital P.

(*CPP* ii. 294–7.)

The play that Shaw began three years after *Caesar and Cleopatra*, *Man and Superman*, is based on this rejection of a progressive historical model. The assumption that underlies *Man and Superman* is that Mankind needs to evolve further in order to cope with the problems of his civilization. In the Epistle Dedicatory, Shaw said that he had no illusions left 'on the subject of education, progress, and so forth', because '[p]rogress can do nothing but make the most of us all as we are' (*CPP* ii. 514). One of the chapters in the Revolutionist's Handbook, entitled 'The Verdict of History', is in part a recapitulation of the arguments we have just been looking at in the *Caesar and Cleopatra* note. In response to the proposition that there has been 'progressive moral evolution operating visibly from grandfather to grandson', Tanner/Shaw replies that:

a thousand years of such evolution would have produced enormous social changes, of which the historical evidence would be overwhelming. But not Macaulay himself, the most confident of Whig meliorists, can produce any such evidence that will bear cross-examination. Compare our conduct and our codes with those mentioned contemporarily in such ancient scriptures and classics as have come down to us, and you will find no jot of ground for the belief that any moral progress whatever has been made in historic time, in spite of all the romantic attempts of historians to reconstruct the past on that assumption.

(*CPP* ii. 773.)[26]

So much for progress with a capital P.

In *Back to Methuselah*, the rejection of the idea of progress is

---

[26] Shaw propounded a similar case against progress in a letter to the editor of the *Daily News* in 1903, the year in which *Man and Superman* was published. Defending a lecture that he had recently given on the subject, he said he did not 'feel disposed to anticipate objections which assume that I am a fool and an ignoramus, and that I have left out of account all the optimistic commonplaces of Macaulay's history' (*Agitations*, 70–2).

Progress an Illusion? 121

evident in the very structure of the work. *Back to Methuselah* consists of five plays, covering history from the Garden of Eden until a period 30,000 years in the future, as far as thought can reach. The present, the time at which the play was written, is placed in the second of the five plays. This is to say that the cycle skips from the Garden of Eden to the present, conveying the conviction that there is nothing of any significance that has occurred in between. The fall that took place in *In the Beginning* is fully felt in *The Gospel of the Brothers Barnabas*, which is set shortly after the First World War. Adam's invention of death, and Cain's invention of murder and war, are thriving.

Tragedy of an Elderly Gentleman, set in the year 3000, has as its title-character a survival from a previous era in the sense that he is a short-liver, most of whose ideas are thoroughly old-fashioned. When he orates to the long-liver Zoo in high Macaulayite rhetoric, he sounds particularly foolish: 'I was illustrating—not, I hope, quite infelicitously—the great march of Progress. I was shewing you how, shortlived as we orientals are, mankind gains in stature from generation to generation, from epoch to epoch, from barbarism to civilization, from civilization to perfection.' In response to Zoo's emphasis on the importance of longevity, he contradicts himself:

THE ELDERLY GENTLEMAN. ... Human nature is human nature, longlived or shortlived, and always will be.
ZOO. Then you give up the idea of progress?
THE ELDERLY GENTLEMAN. I do nothing of the sort. I stand for progress and for freedom broadening down from precedent to precedent.

(Act I, *CPP* v. 517, 520.)

The allusion to Tennyson's poem 'You Ask Me Why' helps to place the Elderly Gentleman's ideas in the Victorian era, just as the Devil's ironic use of Tennyson's lines from *In Memoriam* in the Hell Scene of *Man and Superman* ('Where you now see ... continual ascent by Man on the stepping stones of his dead selves to higher things, you will see nothing but an infinite comedy of illusion') creates a Victorian context for the progressive idea of history that he is contemptuously dismissing. In both cases the Tennysonian attitude is discredited.

When Shaw wrote a new preface for *Fabian Essays in Socialism* in 1908, he noted the danger of revising work one did when younger. '[T]he difference between the view of the young and the elderly is not necessarily a difference between wrong and right. The Tennysonian process of making stepping stones of our dead selves to higher things is pious in intention, but it sometimes leads downstairs instead of up.'[27] That is what happens in Shaw's late attempt along the lines of *Back to Methuselah*, the *Farfetched Fables* that he wrote in 1948. This series of playlets begins with the present, after the dropping of the atomic bomb on Hiroshima, and moves into an unspecified but distant future. In this future the nineteenth and twentieth centuries are seen in retrospect as the nadir of civilization, as in the last play of *Back to Methuselah*. But whereas in *Back to Methuselah* the two throwbacks to our biological epoch, Ozymandias and Cleopatra-Semiramis, are isolated accidents, created by the scientist Pygmalion, in this later play the human race as a whole seems to have relapsed to the Dark Ages of the twentieth century. The characters in the Sixth and Last Fable represent a regression from the future depicted in the fourth and fifth fables. In the Sixth and Last Fable, students in a schoolroom discuss a theory of Disembodied Races, according to which there are higher beings existing as Thought Vortexes. One of the students dislikes the theory, because he 'was brought up to consider that we are the vanguard of civilization, the last step in creative evolution. But according to the theory we are only a survival of the sort of mankind that existed in the twentieth century, no better than black beetles compared to the supermen who, evolved into the disembodied.' Then one of these disembodied ones appears among the students as '*a youth, clothed in feathers like a bird*', and explains that he has become incarnate again because 'Evolution can go backwards as well as forwards' (*CPP* vii. 460, 464).

Farfetched Fables would not be worth much of our attention apart from the fact that it represents, in an extreme form, a current in Shaw's thinking. This idea that evolution will not necessarily carry Mankind upwards expresses itself in a different way in some of Shaw's writing, where he warns that the Life Force may sweep away the human race and replace it with a new creation. 'I see the time coming', said Goethe in the early

---

[27] Shaw, 'Fabian Essays Twenty Years Later', *Essays in Fabian Socialism*, 298.

nineteenth century, 'when God will take no more pleasure in the race, and must again proceed to a rejuvenated creation.'[28] But Goethe placed this possibility in the distant future, while for Shaw it seemed at times as if the human race might soon be reaching the end of its existence. In 1913, for example, he concluded a lecture on 'Christian Economics' with this warning:

> Ladies and gentlemen, it is not an impossible thing that some day or other, there may walk out of a bush somewhere a new being of which you have no conception, not a man at all, or a woman at all; something new that has never occurred before: and the work of God may be handed over to that new thing; and it may be said of us: 'These people have failed; they are scrapped; they are gone.' Part of the mission of the new thing would be to destroy these people as part of the mission of man was to destroy the tiger. Think of that possibility, ladies and gentlemen; and make up your minds to work pretty hard.[29]

This is Shaw's collapse-of-civilizations idea applied in a biological context: not the decline and fall of a civilization but the decline and fall of the human race.

Up to this point I have presented one side of Shaw's attitude towards progress. If this chapter were to end here, it would convey a highly misleading account of its subject, for there is another side to each of the areas we have been considering. Shaw is deeply committed to the idea that there has been no progress in history *so far*, but he has an equally deep commitment to the possibility of progress in the future. As soon as one accepts the future as part of the historian's material, then one has a new perspective that brings Shaw much closer to Macaulay and the Whig, progressivist tradition of Victorian historiography. As long as the historical field is limited to the past several thousand years, Shaw sees a pattern of levelling and recurrence, or even decline, but when the context is expanded to include thousands or millions of years lying behind and also before us, then Shaw sees a pattern of ascent. As a dramatist he sees both patterns at once, as in the Hell Scene of *Man and Superman* where both the Devil's pessimistic theory of recurrence and Juan's optimistic theory of a

---

[28] Quoted in J. B. Bury, *The Idea of Progress* (1920; New York: Dover, 1955), 259.
[29] Shaw, 'On Christian Economics', in Allan Chappelow, *Shaw: 'The Chucker-Out'* (London: George Allen and Unwin, 1969), 154–5.

progressive Life Force are valid and compelling ways of looking at the world. Most of the historical evidence so far is on the side of the Devil, but there is still the consideration that the Serpent offers to Adam and Eve in *In the Beginning*: 'As long as you do not know the future you do not know that it will not be happier than the past. That is hope' (Act I, *CPP* v. 356). It is hope, which means the imagination to conceive the possibility of a better future, that inspires Juan to leave hell for heaven, and it is the absence of hope that underlies the diabolic world-view. 'Written over the gate here are the words, "Leave every hope behind, ye who enter." Only think what a relief that is! For what is hope? A form of moral responsibility. Here there is no hope, and consequently no duty, no work, nothing to be gained by praying, nothing to be lost by doing what you like' (Act III, *CPP* ii. 642). The double perspective in the Hell Scene is found in another form near the beginning of the late play *Buoyant Billions*, where the son tells his father that 'At first sight there is no hope for our civilization', but that 'At second sight the world has a future that will make its people look back on us as a mob of starving savages' (Act I, *CPP* vii. 313). The way that Shaw combines first sight and second sight is an important element in the dramatic vitality of his plays.

The idea that the present is a low point in history also offers a double perspective. If one sees the present as a low point in relation to the past, then one has a theory of decline. If, on the other hand, one shifts the focus and sees the present as a low point in relation to the future, then one has a theory of progress. Instead of an oscillating pattern, or a descending line, one has something like a V-shape, with the present as a trough and the future as an ascending arm. *Back to Methuselah* juxtaposes both of these models of historical development. Like *Paradise Lost*, it is at the same time a work about a fall and a work about ascent. Although the Elderly Gentleman's expression of his belief in progress is made to seem ridiculous, *Back to Methuselah* as a whole is a dramatic enactment of a theory of progress, and indeed he himself is a victim of Mankind's ascent towards greater longevity and greater spiritual power. From the wide perspective created by *Back to Methuselah*, the Elderly Gentleman is quite right in his view that 'mankind gains in stature' in 'the great march of Progress': it is just this gain in stature that is the main subject of the work. In fact, if we look at history in the terms suggested by the time-frame of *Back to Methuselah*, all of the Macaulayite formulations that Shaw scoffs at in passages I have cited in the previous part of this chapter are, *mutatis mutandis*, perfectly valid.

In *Back to Methuselah* Shaw uses the future in the way that Carlyle, Ruskin, and Morris use the past, as a means of revealing the defects of the present. Carlyle's *Past and Present* becomes Shaw's *Future and Present*. Morris's utopian romance *News from Nowhere* also uses the future as an invidious contrast with the present, but Morris's future is a recurrence of an idealized medieval past. The future in *Back to Methuselah*, on the other hand, is not a reversion to a historical past. Although the Arcadian setting of *As Far as Thought Can Reach* could at first suggest an idealized Greece as a model, the life of the Ancients in this play has evolved beyond that of the Garden of Eden in *In the Beginning*. *Back to Methuselah* begins with a decline, within the first two plays (which represent all of history so far), and then proceeds to a recovery of what was lost, in the next play, and finally takes Mankind a stage higher than anything that has been achieved so far, with the intimation of continuing ascent. Whereas in the first play Adam and Eve are children for the most part, the Ancients in the last play are, relatively speaking, adults, and they long for fuller adulthood.

The idea that civilizations collapse and that ours is near the end of its life is another attitude that can point in two directions. Thus far in history civilizations may have succeeded each other in a merely repetitive way, without any overall upward movement, but as long as there is change in history, there is the possibility of improvement. If our civilization is facing a crisis in which greater human capacity is required, then perhaps Mankind will rise to the challenge and achieve greater heights; this is what happens in *Back to Methuselah* after the crisis of the First World War, and this is what Juan predicts in the Hell Scene of *Man and Superman*. And even if our civilization is about to be swept away, then it may be replaced by something better. Or if the human race itself is to be superseded by some new creature, then we may have been replaced by something better. Even the destruction of the human race can be part of a theory of decline or of progress, depending on how far back one is standing. It is a matter of perspective, in that it can be seen as the decline of Mankind, or as the progress of

Life. Shaw sees it both ways. Against the warning that I quoted earlier from the lecture on 'Christian Economics' (above, p. 123) we can set Franklyn Barnabas's speech in *The Gospel of the Brothers Barnabas*, which says the same thing with a different emphasis: 'We shall not be let alone. The force behind evolution, call it what you like, is determined to solve the problem of civilization; and if it cannot do it through us, it will produce more capable agents. You and I are not God's last word: God can still create. If you cannot do His work He will produce some being who can' (*CPP* v. 430).

One play that provides a good example of the Shavian double perspective is *Saint Joan*. We have already seen how this works in terms of the succession of epochs from the Middle Ages to the Renaissance. With respect to the idea of progress, the play is both a dramatization of cyclical recurrence or history as a level line, and at the same time a dramatization of history as ascent. While an audience is aware, as I believe an audience is supposed to be aware, that Joan's individualism and Nationalism are dangerous, we are made to feel at the same time, as we watch the play, that they are desirable. We know that the future which Joan is helping to bring into being will not be an improvement over the past, and yet the play exploits a Macaulayite viewpoint dramatically in making us feel that Joan represents a progressive force in history. Furthermore, there is the fact that Joan is superior to the common run of humanity, and this aspect of the play also supports two divergent readings of historical development. Although she has been defeated in her lifetime we know that she has survived as an inspiration to subsequent generations. But we discover in the Epilogue that the common run of humanity has not improved at all over the past 500 years, and the rejection of Joan in the final scenes of the play is repeated in the timeless world of the Epilogue that extends to the present (with the Gentleman '*in black frock-coat and trousers, and tall hat, in the fashion of the year 1920*' (*CPP* vi. 203)); the suggestion here is one of cyclical recurrence. Nevertheless, Joan's final cry, 'How long, O Lord, how long?' directs our attention to the future, and the end of the Epilogue is no more the end of Joan's story than the end of the Trial scene was. Here, towards the end of Scene vi, Ladvenu says to Warwick, 'This is not the end for her, but the beginning', which prepares us for the concluding lines of dialogue:

THE EXECUTIONER. Her heart would not burn, my lord; but everything that was left is at the bottom of the river. You have heard the last of her.
WARWICK [*with a wry smile, thinking of what Ladvenu said*]. The last of her? Hm! I wonder!

(*CPP* vi. 189, 190.)

Whether we have seen the last of her depends upon our timeframe. In one sense, the crowd at her execution saw the last of her. In another sense, the audience watching the end of the Epilogue has seen the last of her. If we look to the future, however, we can see that the human possibilities she represents could flourish, and the earth *could* become ready to receive its saints. Unless one conceives of this as a possibility, then one's dramatic response to *Saint Joan* has been diminished. The assumption underlying the play is that while there has been no progress in history thus far, there is the possibility of progress in the future.

Saint Joan suggests a theory of progress in quite a different way as well. In many places in Shaw's writing, the main example that he gives in order to demonstrate that the present is no better than the past is the persistence of persecution. Shaw's interest in this subject of persecution and toleration implies both a rejection of progress and a belief in it. If history is a succession of persecutions, and modern toleration is no better than what prevailed in the Middle Ages, then there would be no basis for a belief in progress. But the other side of this issue is that the importance of toleration lies in the relationship between new ideas and human improvement. This is a subject on which Shaw was probably influenced by his reading of Buckle (as well as John Stuart Mill). In recommending Buckle's *History of Civilization* in *The Intelligent Woman's Guide to Socialism*, he summarized its moral: 'that progress depends on the critical people who do not believe everything they are told: that is, on scepticism'.[30] This is an

[30] *Intelligent Woman's Guide*, 501. Here is one of Buckle's passages in praise of unfettered controversy—a passage that brings to mind the Victorian intellectual roots of *Saint Joan*: '[T]he great enemy of knowledge is not error, but inertness. All that we want is discussion, and then we are sure to do well, no matter what our blunders may be. One error conflicts with another; each destroys its opponent, and truth is evolved. This is the course of the human mind, and it is from this point of view that the authors of new ideas, the proposers of new contrivances, and the originators of new heresies, are benefactors of their species. Whether they are right or wrong, is the least part of the question. They tend to excite the mind; they open up the faculties; they stimulate us to

continued overleaf

argument that Shaw used four years earlier in the Preface to *Saint Joan*, and as early as 1898, in *The Perfect Wagnerite*: '[I]f the energy of life is still carrying human nature to higher and higher levels, then the more young people shock their elders and deride and discard their pet institutions the better for the hopes of the world, since the apparent growth of anarchy is only the measure of the rate of improvement.'[31] In the Preface to *Mrs Warren's Profession*, the same argument is part of a denunciation of censorship: 'All censorships exist to prevent anyone from challenging current conceptions and existing institutions. All progress is initiated by challenging current conceptions, and executed by supplanting existing institutions' (*CPP* i. 247). These passages use the favourite Victorian words 'progress' and 'improvement', and in the Preface to *Misalliance* we find this: 'The expediency of Toleration has been forced on us by the fact that progressive enlightenment depends on a fair hearing for doctrines which at first appear seditious, blasphemous, and immoral, and which deeply shock people who never think originally, thought being with them merely a habit and an echo. The deeper ground for Toleration is the nature of creation, which, as we now know, proceeds by evolution' (*CPP* iv. 65). The Preface to *On the Rocks*, which also deals with this subject of toleration, includes a dialogue between Jesus and Pontius Pilate, in which Jesus states the case for toleration in evolutionary terms: 'Without sedition and blasphemy the world would stand still and the Kingdom of God never be a stage nearer. . . . The beast of prey is not striving to return: the kingdom of God is striving to come. The empire that looks back in terror shall give way to the kingdom that looks forward with hope' (*CPP* vi. 623–4).

These are not isolated or uncharacteristic arguments of Shaw's,

---

fresh inquiry; they place old subjects under new aspects; they disturb the public sloth; and they interrupt, rudely, but with most salutary effect, that love of routine, which, by inducing men to go grovelling on in the ways of their ancestors, stands in the path of every improvement, as a constant, an outlying, and, too often, a fatal obstacle' (H. T. Buckle, *History of Civilization in England* (1857–61; 3 vols., London: Longmans, Green, 1878), iii. 394–5 (chap. v)). Another Victorian work that stands behind *Saint Joan* is John Stuart Mill's great essay *On Liberty* (1859), to which Shaw paid tribute in 1933 as his 'own first textbook on the subject' (Preface to *On the Rocks, CPP* vi. 611). The value of toleration is also a significant theme in the writings of Macaulay.

[31] *Shaw's Music*, iii. 482.

but evidence of an important aspect of his world-view. While he vehemently and decisively rejected the idea of progress in history, he at the same time found it highly attractive. Temperamentally he was much nearer to Macaulay's optimistic belief in progress than some of his anti-Macaulayite statements would suggest. One indirect indication of some affinity between the two men might be their common admiration for Bunyan's *Pilgrim's Progress*, which as its title proclaims is a paradigm of ascent. Another more direct indication would be Shaw's attitude towards mechanical improvement. Although he would not accept such improvement as real progress, he was undoubtedly drawn to it. In his second novel, *The Irrational Knot*, written in 1880, it is an unamiable character who declaims against machinery, while the hero of the novel, an American inventor called Edward Conolly, is an agent of moral, social, and industrial progress. Seventy years later, after Shaw had seen, and for the most part responded enthusiastically to, mechanical improvements of the twentieth century, he wrote his last play, *Why She Would Not*. In this work Henry Bossborn shows himself to be the veritable born boss by taking over Serafina White's country house and her business empire, and improving them both by casting out the old architecture and commercial methods, and introducing the latest improvements. When we first see Serafina's '*pretentious country house*', Four Towers, it is '*surrounded by a high stone wall and overshadowed by heavy elm trees. The wall is broken by four sham towers with battlemented tops.*' Its drawing-room is '*overcrowded with massive early Victorian furniture, thick curtains, small but heavy tables crowded with nicnacs, sea shells, stuffed birds in glass cases, carpets and wall paper with huge flower designs, movement obstructed and light excluded in every possible way*'. Then after Bossborn has finished with the place this stuffy, old-fashioned monstrosity has been replaced in the final scene by '*an ultra modern country house dated 1950*', contrasting strongly with its predecessor. Serafina is not happy with the change, but the play makes it clear that it has none the less been a change for the better. Bossborn assures her, 'You could not live in Four Towers now because you are so enormously more comfortable and civilized here' (Scenes ii, iv, v, *CPP* vii. 666, 670, 673, 676).

This play may remind us of Caesar's unsentimental response to the burning of the library at Alexandria:

THEODOTUS [*wildly*]. Will you destroy the past?
CAESAR. Ay, and build the future with its ruins.
(*Caesar and Cleopatra*, Act II, *CPP* ii. 219.)

This Macaulayite contrast between the old and the new is also reflected in a 1904 suggestion (in a discussion of housing) that because outdated houses need to be replaced, 'the municipalities of the future will be almost as active in knocking our towns down as in building them up',[32] and in a glance at the future in the Preface to *Misalliance*: 'Before motor cars became common, necessity had accustomed us to a foulness in our streets which would have horrified us had the street been our drawing-room carpet. Before long we shall be as particular about our streets as we now are about our carpets; and their condition in the nineteenth century will become as forgotten and incredible as the condition of the corridors of palaces and the courts of castles was as late as the eighteenth century' (*CPP* iv. 78). It is interesting to notice just how much like Macaulay Shaw can sound at times.

Shaw's rebellious attitude towards established institutions and ideas is not very much like Macaulay's Burkean sense of a gradually evolving tradition, but Shaw's view that an evolving society outgrows its institutions implies some sort of theory of progress. *The Quintessence of Ibsenism* argues that outdated institutions must be cast aside when they conflict with the growing human will, and the Preface to *Getting Married* speaks of 'a revolt against marriage' which has spread rapidly within the author's recollection. Ordinary, unthinking people, this Preface continues, 'accept social changes today as tamely as their forefathers accepted the Reformation' and subsequent religious changes in England. 'If matters were left to these simple folk, there would never be any changes at all; and society would perish like a snake that could not cast its skins. Nevertheless the snake does change its skin in spite of them; and there are signs that our marriage-law skin is causing discomfort to thoughtful people and will presently be cast whether the others are satisfied with it or not.' There comes a moment when society must change its institutions to keep up with the times. Because of England's delay

---

[32] Shaw, 'The Common Sense of Municipal Trading', *Essays in Fabian Socialism*, 219.

in making the necessary adaptations to its marriage laws, 'Europe and America have left us a century behind in this matter' (*CPP* iii. 459–60, 509).

Another sign of Shaw's temperamental affinity with the idea of progress is his emphasis on keeping up to date. In his music criticism there are many references to outdated taste, music, and ideas, as in his attitude towards the musical culture of Paris: 'I have no patience with Paris: provincialism I do not mind, but a metropolitanism that is fifty years behind the time is insufferable.'[33] Roebuck Ramsden in *Man and Superman* is a comic character in that he imagines himself to be an advanced man while in fact his thinking is thirty or forty years out of date, and in *Everybody's Political What's What?* there is the warning that a civilization can go wrong, by 'falling out-of-date' in its economics, its politics, its science, its education, and its religion. 'I maintain that in all five we are dangerously behind the times, and will go to pieces like all former civilizations known to us unless we give our institutions a thorough overhaul pretty frequently.'[34] Shaw wrote in his preface to the 1931 edition of *Fabian Essays in Socialism* that 'our common education is centuries out of date; and generations of Britons still crowd in on us with a laboriously inculcated stock of ideas of which half belong to the courts of the Plantagenets and the other half to the coffee houses under Queen Anne'.[35] In *On the Rocks*, written two years later, the Prime Minister is a man whose ideas are dangerously out of date, with the result that he finds himself unable to cope with the problems that are confronting the country. Late in the first act he is visited by a Lady who tells him that he is 'a ghost from a very dead past' while she is 'a ghost from the future'. Ghosts from the future, she explains, are 'women and men who are ahead of their time. They alone can lead the present into the future. . . . The ghosts from the past are those who are behind the times, and can only drag the present back' (*CPP* vi. 668–9). Between the two acts she takes the Prime Minister to her retreat in the Welsh mountains where he is brought intellectually up to date.

*

[33] *Shaw's Music*, ii. 886 (1893 article).
[34] *Everybody's Political What's What?*, 344.
[35] Shaw, 'Fabian Essays Forty Years Later: What They Overlooked', *Essays in Fabian Socialism*, 301.

The main reason why Shaw reacted so vehemently to Thomas Tyler's theory of cycles was that Tyler's reading of history was deterministic, and Shaw always rejected determinism in all its forms; the freedom of the human will is fundamental to his world-view. Cycles are part of his reading of history, but there is nothing inevitable in the pattern of recurrence that has dominated historical development thus far. The pattern of history could have been one of continual ascent, had individuals over the centuries made the right choices, and aspired for example to be Caesar rather than to be Cleopatra. There has always been the possibility of progress, and today human beings are faced with the same choices that past generations have had, so that the future is in our hands. Neither the lack of progress so far, nor the reality of progress in the future, is automatic. Nothing in our future is inevitable; nothing is independent of the human will. The true causation is the activity of what Shaw called the evolutionary appetite or the Life Force, the collective will of humanity, which is the sum of individual wills.

This possibility of progress brings us back to Juan's rhetorical question in the Hell Scene of *Man and Superman*: '[H]as the colossal mechanism no purpose?' The answer once again depends upon one's perspective. Within recorded history no purpose is evident, but the debate in the Hell Scene extends beyond the boundaries of history as the term is normally used. Although Shaw regarded himself as radically anti-Darwinian, his intellectual development occurred within the framework of Victorian evolutionary thinking. One of the results of Darwin's *Origin of Species* was to place Nature within the context of history. As R. G. Collingwood put it, *The Origin of Species*

figures as the book which first informed everybody that the old idea of nature as a static system had been abandoned. The effect of this discovery was vastly to increase the prestige of historical thought. Hitherto the relation between historical and scientific thought, i.e. thought about history and thought about nature, had been antagonistic. History demanded for itself a subject-matter essentially progressive; science, one essentially static. With Darwin, the scientific point of view capitulated to the historical, and both now agreed in conceiving their subject-matter as progressive. Evolution could now be used as a generic term covering both historical progress and natural progress. The victory of evolution in scientific circles meant that the positivistic reduction of

Progress an Illusion? 133

history to nature was qualified by a partial reduction of nature to history.[36]

This nineteenth-century revolution in thought enabled Shaw to reverse the old conception of history as progressive and Nature as static. For Shaw, human history has not been exactly static, but it is level. Nature, on the other hand, has had a clear line of ascent from primitive forms to Man, and will continue to ascend in the future, whatever the human race does with itself. Thus Shaw is able to give expression to his Macaulayite temperament in a theory of history, in which history is a matter of biology rather than mere politics or social institutions. Thus *Man and Superman*, with all of the denunciations of the progressive model in the Epistle Dedicatory, the Revolutionist's Handbook, and the Devil's speeches in the Hell Scene, is ultimately a dramatic study of the aspiring universal will driving life to higher and higher levels. And thus *Back to Methuselah*, with all of its suggestions of cyclical recurrence, offers this insight of Franklyn Barnabas right after the remark about Savvy as a new hat and frock on Eve:

FRANKLYN. Yes. Bodies and minds ever better and better fitted to carry out Its [Life's] eternal pursuit.
LUBIN [*with quiet scepticism*]. What pursuit, may one ask, Mr Barnabas?
FRANKLYN. The pursuit of omnipotence and omniscience. Greater power and greater knowledge: these are what we are all pursuing even at the risk of our lives and the sacrifice of our pleasures. Evolution is that pursuit and nothing else. It is the path to godhead. A man differs from a microbe only in being further on the path.
(*The Gospel of the Brothers Barnabas*, CPP v. 423.)

Or as Don Juan declares in the Hell Scene of *Man and Superman*: 'Life is a force which has made innumerable experiments in organizing itself ... the mammoth and the man, the mouse and the megatherium, the flies and the fleas and the Fathers of the

---

[36] R. G. Collingwood, *The Idea of History* (1946; London: Oxford Univ. Press, 1977), 129. This was a development that occurred in the middle of the nineteenth century, but it did not entirely await the publication of *The Origin of Species* in 1859; Tennyson's *In Memoriam*, published nine years earlier, reconciles the poet's anxiety and hope by reaching a conception of Nature as cruel in the short run but ultimately progressive. *Man and Superman* may quote Tennyson's poem derisively, but the two works are by no means dissimilar in their theories of history—a nice irony in view of *Man and Superman*'s emphasis on being intellectually up to date.

Church, are all more or less successful attempts to build up that raw force into higher and higher individuals, the ideal individual being omnipotent, omniscient, infallible, and withal completely, unilludedly self-conscious: in short, a god' (Act III, *CPP* ii. 661–2). When the cycles are seen in a wide biological context rather than the narrow context of traditional history, they turn out to be a spiral; recurrence ultimately turns out to be ascent.

'[T]he history of our country during the last hundred and sixty years is eminently the history of physical, of moral, and of intellectual improvement', asserted Macaulay.[37] For Juan in the Hell Scene, the history of Life over the last several million years is eminently the history of physical, of moral, of intellectual improvement. For Macaulay, Victorian England is Mankind's highest achievement so far. For Shaw, Mankind is Nature's highest achievement so far. The difference is that while Macaulay acknowledged the progress that would be made beyond the present level, his comparisons between past and present led him to admire what had now been achieved. In Shaw's case, on the other hand, comparisons between the present and a possible future led him to a profound dissatisfaction with what had now been achieved. 'I tell you', says Juan, 'that as long as I can conceive something better than myself I cannot be easy unless I am striving to bring it into existence or clearing the way for it. That is the law of my life' (*Man and Superman*, Act III, *CPP* ii. 679–80). Mankind has made no progress within historical time, but within the biological, evolutionary context the progress from the most primitive forms of life has been considerable. The result so far—Man as he is at present—is not satisfactory, but there is no end to the progress that is possible in the future.

---

[37] Macaulay, *History of England*, *Works*, i. 2 (chap. 1).

5

Present History

> O ye hypocrites, ye can discern the face of the sky; but can ye not discern the signs of the times?
>
> (Matthew 16: 3.)
>
> [I]t seems to me that so great power ought not to be spent on visions of things past but on the living present.
>
> (Ruskin to Tennyson, on *Idylls of the King*.*)

The fundamental satirical strategy of *Back to Methuselah* is revealed in Zoo's reply to the Elderly Gentleman's request that he be assigned to the care of one of the long-livers' offspring who have reverted to the ancestral type and were born short-lived:

THE ELDERLY GENTLEMAN [*eagerly*]. . . . I hope you will not be offended if I say that it would be a great comfort to me if I could be placed in charge of one of those normal individuals.

ZOO. Abnormal, you mean. What you ask is impossible: we weed them all out.

(*CPP* v. 528.)

Like much of Shaw's work, *Back to Methuselah* seeks to destroy our concept of normality, to alter our perspective so that we see ourselves from the outside, deprived of our conventional assumptions about our central, natural place in the scheme of things. Sometimes, it is the geographical perspective that is altered, as in Cleopatra's question to Britannus: 'Is it true that when Caesar caught you on that island, you were painted all over blue?' (*Caesar and Cleopatra*, Act II, *CPP* ii. 222). England (or Britain) is no longer the centre of the world, but merely 'that island', or as Caesar describes it from his Roman perspective in Act III, 'that western land of romance—the last piece of earth on the edge of the ocean that surrounds the world' (*CPP* ii. 268–9). Lines such as these are intended to make an English audience see themselves

---

* [Hallam Tennyson], *Alfred Lord Tennyson: A Memoir* (2 vols., London: Macmillan, 1897), i. 453.

from an uncomfortable perspective that moves them from the centre to the periphery. Such is the effect, too, of Captain Shotover's more direct challenge in *Heartbreak House*: 'Do you think the laws of God will be suspended in favor of England because you were born in it?' (*CPP* v. 177). The import of Shotover's question, as of Cleopatra's, is that England is just another country, with no special position in the world.

Cleopatra's question shifts the centre of the world from England to Egypt. In *Back to Methuselah* this shifting of the centre from west to east is suggested in the final speech of *The Thing Happens* when Confucius, the Chinese civil servant who seems to run the British Islands in AD 2170, exclaims, 'Oh, these English! these crude young civilizations!' (*CPP* v. 490). The first act of the play that follows is set in Galway in the year 3000. The Elderly Gentleman meets a local woman:

THE WOMAN. You are a foreigner, are you not?
THE ELDERLY GENTLEMAN. No. You must not regard me as a foreigner. I am a Briton.
THE WOMAN. You come from some part of the British Commonwealth?
THE ELDERLY GENTLEMAN [*amiably pompous*]. From its capital, madam.
THE WOMAN. From Baghdad?
THE ELDERLY GENTLEMAN. Yes. You may not be aware, madam, that these islands were once the centre of the British Commonwealth, during a period now known as The Exile. They were its headquarters a thousand years ago.

(*Tragedy of an Elderly Gentleman*, *CPP* v. 492.)

The centre has apparently been reduced to the periphery (although the real power, it turns out, lies with the long-lived inhabitants of what used to be the British Isles). Similarly, in *The Simpleton of the Unexpected Isles*, which is set at some point in the future, a character remarks that 'since India won Dominion status Delhi has been the centre of the British Empire' (Act II, *CPP* vi. 807).

These passages bring to mind the emperor Constantine's transfer of the capital of the Roman Empire from Rome to Constantinople in the fourth century, and they reflect Shaw's own perception of what had happened in more recent history, and what could happen in the future. Capitalism, he wrote in *The Intelligent Woman's Guide to Socialism*, caused England to acquire an empire, 'with the curious result, quite unintended by the British people, that the centre of the British Empire is now in the East instead of in Great Britain, and out of every hundred of our fellow subjects only eleven are whites, or even Christians'. This shift from west to east could take a more tangible form: one could suddenly discover that the Houses of Parliament and the King had moved to Constantinople or Baghdad or Zanzibar, 'and that this insignificant island is to be retained only as a meteorological station, a bird sanctuary, and a place of pilgrimage for American tourists'. Such an event 'would be a perfectly logical development of Capitalism. And it is no more impossible than the transfer of the mighty Roman empire from Rome to Constantinople was impossible.'[1]

Shaw's work tries to challenge our normal sense of place, and also our normal sense of historical time. Just as it offers a new perspective from the point of view of Alexandria or Baghdad, so it offers a new perspective from the point of view of AD 31,920 or some other period in the future. Writing in 1929 about the events in Ireland leading to independence, Shaw said that 'Future historians will probably see in these catastrophes a ritual of human sacrifice without which the savages of the twentieth century could not effect any redistribution of political power or wealth' (Preface to *John Bull's Other Island* ('Twentyfour Years Later'), *CPP* ii. 889). In *As Far as Thought Can Reach*, the youthful scientist Pygmalion in 31,920 is telling his comrades about a time in the past when people suffered from cancer, and one of his listeners admonishes him to '[k]eep off the primitive tribes' (*CPP* v. 595). *Back to Methuselah* plays with historical perspective in order to show what the present might look like when seen from outside —from the vantage-point of the future. Such a work makes us see ourselves as history may come to see us, or in fact as Whig historians see the Middle Ages. The present loses its special place in history, and becomes just another point in the historical continuum. This attitude of Shaw's partly (but not entirely) explains some of his notorious statements in the 1920s and 1930s about Fascist dictators. Seeing the present as just another historical period, he treated current affairs with the

---

[1] Shaw, *The Intelligent Woman's Guide to Socialism, Capitalism, Sovietism and Fascism* (London: Constable, 1949), 313–14.

detachment of the historian, and in the way that a provocative historian might find something good to say about Nero or Attila, Shaw found something good to say about Mussolini and Hitler.

Back to Methuselah shakes our usual notions about time in other ways as well. The opening stage direction reads, '*The Garden of Eden. Afternoon*', while the second act of *In the Beginning* opens with this: '*A few centuries later. Morning*' (*CPP* v. 340, 360), much as the second act of another play might be set 'a few days later'. The passage of vast stretches of time is reflected in the confusion about names from the past, like that of the composite historian Thucyderodotus Macollybuckle. In *As Far as Thought Can Reach*, Pygmalion talks about 'a biologist who extracted certain unspecified minerals from the earth and ... "breathed into their nostrils the breath of life"'. Not much is known about this ancient figure. 'There are some fragments of pictures and documents which represent him as walking in a garden and advising people to cultivate their gardens. His name has come down to us in several forms. One of them is Jove. Another is Voltaire' (*CPP* v. 590–1). The passage of time has eroded the distinction—a rather clear one to those of us living in the twentieth century—between the time recorded in the Book of Genesis and the time of Voltaire. We have been pulled loose from the historical framework that is so familiar to us, as we are in other ways in the Hell Scene of *Man and Superman* and the Epilogue to *Saint Joan*, which take us into a trans-historical world beyond the usual constraints of time.

Shaw's work, then, deprives the present of any special status, just as it deprives England of any special status. The place and time in which an audience happen to find themselves have no sacred position in the scheme of things. England is just another country, and the present is just another historical period. The idea of the present as a historical period no different fundamentally from a past or future era is an important conception to bear in mind in approaching a number of Shaw's plays. Whereas in his history plays he treats the past as if it were the present, so in many of his plays with contemporary settings he treats the present as a period of history. As we saw in the last chapter, by anachronisms and other such means the history plays keep our attention directed towards both past and present. We shall see in this chapter how by other means some of Shaw's plays about the present (including a number of his best plays) keep our attention directed towards both present and past. The history plays make us see our own world reflected in the past;[2] the 'plays present' make us see our own world as emerging from the past. The assumption in both of these cases is one that we have encountered already: that there is no fundamental distinction between past and present. Therefore Shaw's prose can leap over historical time, and therefore his plays can illuminate the past by placing it in a present context, and the present by placing it in a historical context. Shaw would have agreed up to a point with the aphorism of the Victorian historian Edward Freeman that 'History is past politics, and politics is present history'[3]—up to a point, because history for him involved more than politics, but he would have agreed with the equation of past and present implied in Freeman's statement.

Shaw's sense of 'present history' is evident in his treatment of Wagner's *Ring of the Nibelung* in *The Perfect Wagnerite*. In his 1913 Preface to the Third Edition of this book, he explained that the 'centres of gravity' in *The Ring* are missed by many opera-goers. Much of what is central to *The Ring*, 'obvious as it was to Wagner, and as it is to anyone who has reflected on human history and destiny in the light of a competent knowledge of modern capitalistic civilization, is an absolute blank to many persons who are highly susceptible to the musical qualities of Wagner's music and poetry, but have never reflected on human destiny at all, and have been brought up in polite ignorance of the infernal depths our human society descended to in the XIX century'. It is on account of this widespread ignorance that Shaw decided to add to *The Perfect Wagnerite* 'a chapter dealing neither with music nor poetry, but with European history. For it was in that massive material, and not in mere crotchets and quavers, that Wagner found the stuff for his masterpiece.' At the conclusion of his discussion of *Das Rheingold*, Shaw observed that it is the least popular of the four parts of *The Ring* because

---

[2] Such reflections of the present in the historical past are the main subject of A. Dwight Culler's study, *The Victorian Mirror of History* (New Haven and London: Yale Univ. Press, 1985), which examines historical attitudes of Macaulay, Mill, Carlyle, Matthew Arnold, Ruskin, and others.
[3] Quoted in J. W. Burrow, *A Liberal Descent: Victorian Historians and the English Past* (Cambridge: Cambridge Univ. Press, 1981), 163–4.

most people do not understand its real subjects. 'Only those of wider consciousness can follow it breathlessly, seeing in it the whole tragedy of human history and the whole horror of the dilemmas from which the world is shrinking today.'[4] Shaw interprets *The Ring* as a study of major forces in nineteenth-century European history, and he sees the characters as allegorical representations of these historical forces. Moreover, the history that forms the basis of *The Ring* is present history, the contemporary developments in Europe during Wagner's lifetime. The very first point that Shaw made about the work in *The Perfect Wagnerite* was that 'The Ring, with all its gods and giants and dwarfs, its water-maidens and Valkyries, its wishing-cap, magic ring, enchanted sword, and miraculous treasure, is a drama of today, and not of a remote and fabulous antiquity. It could not have been written before the second half of the XIX century, because it deals with events which were only then consummating themselves.' Wagner himself never faced the fact that his work had taken him beyond the confines of the Nibelung epic and that it 'really demanded modern costumes, tall hats for Tarnhelms, factories for Nibelheims, villas for Valhallas, and so on'.[5]

This is the kind of work that Shaw undertakes in plays like *John Bull's Other Island* and *Major Barbara*; he writes historical drama in modern dress, drama with a contemporary setting that reveals the essential historical forces of the present time. 'My business', he told a friend in 1889, 'is to incarnate the Zeitgeist',[6] a phrase which could stand as an apt epigraph for his plays and indeed all of his writing. The mission of revealing to one's contemporaries the spirit of the age has a distinguished Victorian lineage behind it. John Stuart Mill wrote in 1831, in his essay entitled 'The Spirit of the Age', that one's own age 'also is history, and the most important part of history, and the only part which a man may know and understand, with absolute certainty',[7] and he described some of the leading characteristics of English government and society in the early 1830s, in order to make his readers

[4] *Shaw's Music*, ed. Dan H. Laurence (3 vols., London: Max Reinhardt, The Bodley Head, 1981), iii. 414–15, 441–2.
[5] *Shaw's Music*, iii. 421, 493.
[6] Shaw to Tighe Hopkins, 31 Aug. 1889, *Collected Letters 1874–1897*, ed. Dan H. Laurence (London: Max Reinhardt, 1965), 222.
[7] J. S. Mill, 'The Spirit of the Age', *Essays on Politics and Culture*, ed. Gertrude Himmelfarb (Gloucester, Mass.: Peter Smith, 1973), 3.

aware that those who governed were not the best qualified to do so. In the 1840s Disraeli published his trilogy of *Coningsby*, *Sybil*, and *Tancred*, which set out in fictional form his analysis of the problems and possibilities that faced England at that time. He said that in *Coningsby* he had sought, '[i]n an age of political infidelity', to impress 'upon the rising race not to despair, but to seek in a right understanding of the history of their country and in the energies of heroic youth—the elements of national welfare'. *Sybil*, he wrote, 'advances another step in the same emprise. From the state of parties it now would draw public thought to the state of the People whom those parties for two centuries have governed. The comprehension and the cure of this greater theme depend upon the same agencies as the first: it is the past alone that can explain the present, and it is youth that alone can mould the remedial future.'[8] Disraeli's novels promulgate a revisionist, Tory reading of English history—in which, for example, the 1688 Revolution is disastrous rather than glorious—so that his readers will reach a proper understanding of the present through a correction of their Whig misunderstandings about the past.

A good term to describe Disraeli's trilogy would be one derived from Carlyle's 1839 essay, 'Chartism', in which he objected to the fact that Parliament pursued such irrelevancies as the Canada question and the Queen's Bedchamber question while ignoring the central issue of the day: 'the Condition-of-England question'.[9] It is the condition-of-England question that Disraeli takes up in novels like *Coningsby* and *Sybil*, which offer an anatomy of his society at the present time. This is what Carlyle, too, does in much of his writing on contemporary England. The title of one of his early essays is 'Signs of the Times', and the first of his *Latter-Day Pamphlets* is entitled 'The Present Time'. One of the best examples of his writing on the condition-of-England question is 'Shooting Niagara: And After?', published in 1867 during the controversy over the Second Reform Bill and violently protesting against an extension of the franchise. It is antithetical to Mill's 'The Spirit of the Age' in its stand on parliamentary

---

[8] Benjamin Disraeli, *Sybil, or The Two Nations* (World's Classics, London: Oxford Univ. Press, 1975), 429–30 (Book vi, chap. XIII).
[9] Thomas Carlyle, 'Chartism', *Critical and Miscellaneous Essays*, iv. 121, *The Works of Thomas Carlyle*, ed. H. D. Traill (Centenary Edn., 30 vols., London: Chapman and Hall, 1896–9), xxix.

reform, but both essays look at the question against a wide background of the state of their society at the time. Another condition-of-England study that was done in response to the Second Reform Bill controversy was Matthew Arnold's *Culture and Anarchy*, and all of the works I have been citing stand behind such Edwardian novels as H. G. Wells's *Tono-Bungay* and E. M. Forster's *Howards End*. Most of these works are concerned with relationships between social classes, and many of them use representative characters to signify particular types of contemporary Englishmen. All of these works investigate the nature of English civilization at the present time, attempting to discern the signs of the times and the place of the contemporary world in the historical process.

Carlyle began 'The Present Time' with this warning:

The Present Time, youngest-born of Eternity, child and heir of all the Past Times with their good and evil, and parent of all the Future, is ever a 'New Era' to the thinking man; and comes with new questions and significance, however commonplace it look: to know *it*, and what it bids us do, is ever the sum of knowledge for all of us. This new Day, sent us out of Heaven, this also has its heavenly omens;—amid the bustling trivialities and loud empty noises, its silent monitions, which, if we cannot read and obey, it will not be well with us! No;—nor is there any sin more fearfully avenged on men and Nations than that same, which indeed includes and presupposes all manner of sins: the sin which our old pious fathers called 'judicial blindness';—which we, with our light habits, may still call misinterpretation of the Time that now is; disloyalty to its real meanings and monitions, stupid disregard of these, stupid adherence active or passive to the counterfeits and mere current semblances of these. This is true of all times and days.[10]

It is well, I believe, to read most of Shaw's plays about the present time in the light of this passage of Carlyle's, and in the light of all the Victorian and Edwardian works I have touched on in previous paragraphs. Shaw felt that there was a need for condition-of-England plays, for plays that would dislodge fatal 'misinterpretation of the Time that now is'. In his 'Author's Apology' for one of his plays about the past, *Great Catherine*, he said that neither Catherine nor the statesmen with whom she dealt 'had any notion

[10] Carlyle, 'The Present Time', *Latter-Day Pamphlets*, Works, xx. 1.

of the real history of their own times' (*CPP* iv. 897), and his own fictional monarch of the near future, King Magnus in *The Apple Cart*, complains that 'political science, the science by which civilization must live or die, is busy explaining the past whilst we have to grapple with the present: it leaves the ground before our feet in black darkness whilst it lights up every corner of the landscape behind us' (Act I, *CPP* vi. 324). One of the functions of Shaw's plays is to light up the ground before our feet, to reveal and explain the signs of the times so that we can see where we are and where we are going.

Such an intention is suggested in a humorous way in a self-drafted interview published shortly before the opening of Shaw's first play, *Widowers' Houses*, in 1892:

'And is the play to last five hours, if I may ask?' [questioned the 'interviewer'].

'No. There is only time to learn three acts of it, which will occupy no more than the usual time.'

'How many acts, then, are there in the whole play?'

'Seventeen. "Widowers' Houses" is a mere episode in a historic drama.' (Here Mr. Shaw held me spellbound for nearly an hour with a brilliant aperçu of the social and industrial development of England from the Reformation up to the twenty-second century, of which he has the clearest prevision.)

('The Playwright on His First Play', *CPP* i. 126.)

The same point is made in a more serious fashion in Shaw's description of his three Unpleasant Plays as 'criticisms of a special phase, the capitalist phase, of modern social organization';[11] and one ought to look at *Widower's Houses* and *Mrs Warren's Profession* as plays that put the condition-of-England question on the stage. Characters like Sartorius the slum landlord and Mrs Warren the brothel-keeper are representatives of economic and social forces that predominate at the present time. These two plays, like many of Shaw's others, express in dramatic form the dynamics of the present historical epoch.

I would say that almost half of Shaw's plays are concerned with 'present history', in that they treat the present as a historical period, dramatize the main currents of our civilization, and cause

---

[11] Shaw to R. Golding Bright, 10 June 1896, *Collected Letters 1874–1897*, 632.

us to see the contemporary world in relation to a historical background on the implicit assumption that 'it is the past alone that can explain the present'. Some plays conform only in a minor way to this description, but others—most notably *John Bull's Other Island*, *Major Barbara*, and *Heartbreak House*—cannot be properly considered outside such a context.

Good examples of the 'present history' play are three of Shaw's Edwardian works: *Getting Married* (1907–8), *Misalliance* (1909), and *Fanny's First Play* (1910–11). In *Getting Married*, Lesbia describes herself as 'a glorious strong-minded old maid of old England', and refuses to marry because she finds the present legal conditions of marriage intolerable. She protests that 'Just because I have the qualities my country wants most I shall go barren to my grave; whilst the women who have neither the strength to resist marriage nor the intelligence to understand its infinite dishonor will make the England of the future' (*CPP* iii. 652, 634). This concern with 'the England of the future' has the ring of the Edwardian condition-of-England work about it: the ring of *Tono-Bungay* and *Howards End*, for example. The play has characters who represent the Church, the Army, municipal government, and tradesmen (rather in the way that the Church, the Aristocracy, the Army, and the Throne are represented in Scenes iv and v of *Saint Joan*), and all of this is placed against a background of the past. The setting is a medieval episcopal palace; one of the characters declares his admiration for Mahomet; and the Bishop, who is writing a history of marriage, is currently engaged on his chapter on the marriage customs of ancient Rome.[12] Shaw advised an actor who was to play the part of the Bishop that his family group must be played seriously: '[T]he conception of the old English Bridgenorth family is an essential part of the play.'[13] The name of the Bishop's wife, Alice Bridgenorth, is from Sir Walter Scott's historical novel about the Restoration period, *Peveril of the Peak*, and the name is also an

---

[12] The original of the Bishop was Mandell Creighton, the Bishop of London (see Shaw to William Faversham, 3 Sept. 1916, *Collected Letters 1911–1925*, ed. Dan H. Laurence (London: Max Reinhardt, 1985), 412). Creighton had been an academic historian, who wrote a *History of the Papacy* and served as the first editor of the *English Historical Review*; see John Kenyon, *The History Men* (London: Weidenfeld and Nicolson, 1983), 133–5, 166–7, 191–3.

[13] Shaw to William Faversham, 14 Jan. 1916, *Collected Letters 1911–1925*, 349.

echo from one of Shakespeare's history plays: Bridgenorth is the place where Henry IV's armies assemble in Act III of *1 Henry IV* (see III. ii. 174–5). Another name that brings English history to mind is that of Oliver Cromwell Soames, the high church ritualist son of a Noncomformist divine. The inappropriately named Soames agrees with another character that 'many of our landed estates were stolen from the Church by Henry the eighth' (*CPP* iii. 660–1).

Misalliance is very much a contemporary social history play. It dramatizes the essentials of English society in 1909, and as in *Getting Married* the characters are clearly representative of social groups—Tarleton, of the prosperous mercantile middle class, and Summerhays, of the aristocracy, for example. In its attention to relationships between classes it is reminiscent of Arnold's *Culture and Anarchy* (from which Shaw adapts a sentence in his Preface).[14] Like *Culture and Anarchy*—and also like *Howards End*, which was published in the year that the play was first produced—it asks who is worthy to inherit England. Tarleton's son Johnny, like Charles Wilcox in Forster's novel and the Philistine middle class in Arnold's study, believes that 'the time has come for sane, healthy, unpretending men like me to make a stand against this conspiracy of the writing and talking and artistic lot to put us in the back row.... It's we that run the country for them; and all the thanks we get is to be told we're Philistines and vulgar tradesmen and sordid city men and so forth ...' (*CPP* iv. 171).

Like *Culture and Anarchy*, *Misalliance* examines the various social classes to reveal that no existing group is worthy to take over the country, or capable of doing so. The character who comes the closest to seeing this is the ridiculous clerk Gunner: 'Oh! is Hypatia your daughter? And Joey is *Mister* Percival, is he? One of your set, I suppose. One of the smart set! One of the

---

[14] Shaw: '[T]his right to live includes, and in fact is, the right to be what the child likes and can, to do what it likes and can, to make what it likes and can, to think what it likes and can, to smash what it dislikes and can ...' (Preface to *Misalliance*, *CPP* iv. 51. Arnold: '[T]his and that man, and this and that body of men, all over the country, are beginning to assert and put in practice an Englishman's right to do what he likes; his right to march where he likes, meet where he likes, enter where he likes, hoot as he likes, threaten as he likes, smash as he likes' (*Culture and Anarchy*, ed. R. H. Super (Ann Arbor: Univ. of Michigan Press, 1965), 119, chap. II, 'Doing as One Likes'). Note the contrary attitudes towards anarchic behaviour.

bridge-playing, eighty-horse-power, week-ender set! One of the johnnies I slave for! Well, Joey has more decency than your daughter, anyhow. The women are the worst. I never believed it til I saw it with my own eyes. Well, it wont last for ever. The writing is on the wall. Rome fell. Babylon fell. Hindhead's turn will come' (*CPP* iv. 219). This speech is ludicrous, but at the same time an expression of the historical insights of the play. It is not only middle-class Hindhead's turn, but that of all the social classes in the country, including the lower middle class that Gunner himself represents. The ruling classes of the nineteenth century, the aristocracy and the capitalists, have degenerated into the effete Bentley and the thick-headed Johnny. The effective vitality in the play—the vitality that will take over the country —lies with the outsiders, the half-Italian Joey and the Polish Lina, who are victorious in the mating plot and carry off Hypatia and Bentley. In *Misalliance* we have social comedy that reveals the directions of history. The play glances backwards in its inclusion of Gunner's excuse that 'Frederick the Great ran away from a battle' (*CCP* iv. 235), and in Hypatia's name, the original Hypatia having been a Neoplatonist philosopher in fifth-century Alexandria and the subject of a Victorian historical novel by Charles Kingsley (which brings in the threat to the Roman Empire posed by the Goths); and *Misalliance* makes a point of being right up to date in putting an aeroplane crash on the stage in 1910—as *Candida* displayed a typewriter in 1894 and *Man and Superman* an automobile in 1901–2.

'[T]he shutters are up on the gentlemanly business,' says Tarleton in *Misalliance* (*CPP* iv. 229). This is the main theme of *Fanny's First Play*, which also exposes the bankruptcy of contemporary English society. Even if it is 'but a potboiler' (as Shaw described it in his Preface (*CPP* iv. 345)), it like *Misalliance* sounds in a muted way the apocalyptic chord that dominates *Heartbreak House*. Gilbey's question, 'What is the world coming to?,' is more than a banal turn of phrase. Like Gunner in *Misalliance*, he unwittingly gives expression to the issue that is at the heart of the play.[15] The respectable middle-class characters

---

[15] Cf. the position of Charles Lomax in *Major Barbara*: Shaw told Gilbert Murray that 'The moral is drawn by Lomax "There is a certain amount of tosh about this notion of wickedness"' (7 Oct. 1905, *Collected Letters 1898–1910*, ed. Dan H. Laurence (London: Max Reinhardt, 1972), 566).

mention earthquakes in San Francisco, Jamaica, Martinique, and Messina, and plague in China, floods in France, and then:

GILBEY. My Bobby in Wormwood Scrubs!
KNOX. Margaret in Holloway!
GILBEY. And now my footman tells me his brother's a duke!
KNOX. } {No!
MRS KNOX.} {Whats that?
GILBEY. Just before he let you in. A duke! Here has everything been respectable from the beginning of the world, as you may say, to the present day; and all of a sudden everything is turned upside down.
MRS KNOX. It's like in the book of Revelations.

(*CPP* iv. 414–15.)

'What I want to know is, whats to be the end of this?' (*CPP* iv. 426), says Knox (who bears another of the historical names that Shaw chooses for characters in the 'present history' plays). He is speaking about the personal entanglements of the plot, but his question applies to the whole society of which he is a part. As in *Misalliance*, it is the outsiders who carry off the young people of the established social classes: the winners are the self-declassed aristocrat and the loose Darling Dora, both of whom are altogether outside the regular social routine. The 'end of this' will be the disintegration of the present society that we regard as a fact of nature. *Fanny's First Play* shows the way that Fanny O'Dowda, herself a sign of the times as a member of the Cambridge University Fabian Society, sees her contemporary world, and it shows the way that the dramatist who created her writes plays with an eye on the historical process.

Other examples of the 'present history' play would be three works which put contemporary political figures on the stage and dramatize their attempts to cope with the problems of the day: *The Gospel of the Brothers Barnabas* (1918), *On The Rocks* (1933), and *Geneva* (1936). (And one might also mention Shaw's 1909 one-act suffragette play, *Press Cuttings*, which is set in the '*forenoon of the first of April, three years hence*' (*CPP* iii. 840)). The most satirical treatment is in *The Gospel of the Brothers Barnabas*, in which the two politicians, clearly modelled on Asquith and Lloyd George, show themselves utterly incapable of dealing with the problems that led to the War. They are reminiscent of Arnold's lampooning use of contemporary political figures (such as the Radical Member of Parliament, John Arthur

Roebuck), and of Carlyle's fictional politicians like Sir Jabesh Windbag in *Past and Present*. Burge and Lubin reveal the political bankruptcy of the present time, and the need to look beyond Parliamentary politics to find a way to avoid the total collapse of civilization and the possible extinction of the human race. The mindlessness of those who are in power at present will result in a state of anarchy, and the play's warnings are even more dire and far-reaching than those of Carlyle or Arnold. The play is a response to the First World War, as 'Chartism' is a response to the Chartist agitations of the late 1830s, and *Culture and Anarchy* a response to the Second Reform Bill agitations of the 1860s. Like the other 'present history' plays under discussion in this chapter, *The Gospel of the Brothers Barnabas* shows us where we are at the present time; and, on account of its position within the *Back to Methuselah* cycle, it makes explicit what is implicit in other Shavian plays about the present: that the present is a part of a continuing historical process.

On the Rocks also dramatizes the contemporary political scene, and it places current issues in a historical context. The historical dimension of the play is suggested, for example, by the portrait of Walpole beneath which the Prime Minister, Sir Arthur Chavender, is reading *The Times* in the Cabinet Room at Number 10, Downing Street as the play opens (*CPP* vi. 629). In the second act, the working-class Alderwoman Aloysia Brollikins, who comes to Downing Street as part of a delegation, tells the Duke of Domesday how his ancestors 'drove a whole countryside of honest hardworking Scotch crofters into the sea, and turned their little farms into deer forests because you could get more shooting rents out of them in that way'. This story, she says, is not to be found in 'your school histories; but in the new histories, the histories of the proletariat, it has been written, not by the venal academic triflers you call historians, but by the prophets of the new order'. She also tells the Duke that his class will never give up its privileges voluntarily. 'History teaches us that: the history you never read', to which he replies with his own arguments drawn from the past: 'I was merely going to point out, as between one student of history and another, that in the French Revolution it was the nobility who voluntarily abolished all their own privileges at a single sitting, on the sentimental principles they had acquired from reading the works of Karl Marx's revolutionary predecessor Rousseau.' He declares that 'That bit of history is repeating itself today', in that he and other representatives of established classes are prepared to accept the Prime Minister's revolutionary programme. Other characters in this discussion refer to Cromwell's execution of Charles I, and to Guy Fawkes as '[t]he only man that ever had a proper understanding of Parliament' (Act II, *CPP* vi. 696–7, 700–1, 707, 717).

While past history is being used to support characters' arguments, current history is being made in the Cabinet Room as we see characters representing all aspects of contemporary English life discussing what is to be done in the face of the Depression. *On the Rocks* is one of Shaw's *Latter-Day Pamphlets*. The title of Carlyle's first pamphlet, 'The Present Time', would apply perfectly to the play, and the title of another of the *Latter-Day Pamphlets*, 'Downing Street', would do very nicely as well. Indeed, Shaw might have had 'The Present Time' in mind in writing this piece of present history, for Carlyle's pamphlet concludes with a '[s]*peech of the British Prime-Minister to the floods of Irish and other Beggars, the able-bodied Lackalls, nomadic or stationary, and the general assembly, outdoor and indoor, of the Pauper Populations of these Realms*',[16] while *On the Rocks* concludes with the Prime Minister hearing the sounds of a demonstration of the unemployed outside on the street in front of his residence.

Geneva, in its first edition (1939), was subtitled 'a fancied page of history in three acts'.[17] Like *On the Rocks*, it dramatizes the current political crisis, which in this play is the rise of Fascism in Europe. Hitler, Mussolini, and Franco, under slightly disguised stage-names, appear before the League of Nations to argue their respective cases. Among the other characters are a Jewish victim of 'Battler's' anti-Semitic policies, an English bishop, and a Soviet commissar. Once again we see the historical forces of the present, as they express themselves through conflict between characters who represent various fundamental social, political, or national groupings—in this case in the whole of Europe rather than just in England.

Another fancied page of history is *The Apple Cart*. This is not

[16] Carlyle, 'The Present Time', *Latter-Day Pamphlets*, *Works*, xx. 38.
[17] Dan H. Laurence, *Bernard Shaw: A Bibliography* (2 vols., Oxford: Clarendon Press, 1983), i. 232.

exactly a play about present history, in that it is set in the future; but the play's politicians are based on the Ramsay MacDonald government that was in office when the work was first staged and published (1929). At the time of its first English production Shaw informed an interviewer that it was a work of such seriousness 'that I intend to tell Mr. MacDonald when he returns from Geneva that he must refuse to take any young man into his Cabinet who hasn't seen "The Apple Cart" at least six times. It is intended as a salutary lesson, as I feel it is a state of things into which we could drift' (*CPP* vi. 381). It is an extrapolation of present affairs into a period a generation or two in the future. The play reveals the real locus of power in England in the 1920s: Capitalism, as represented by Breakages, Limited; and the real position of England in the world: a maker of confectionery for Americans. Thus *The Apple Cart* is concerned not only with future history, but with present history as well.

*

Shaw wrote in his 1912 Preface to *John Bull's Other Island* that political changes subsequent to the composition of the play meant 'that John Bull's Other Island, which had up to that moment been a topical play, immediately became a historical one' (*CPP* ii. 875). But *John Bull's Other Island* is a historical play not only in this sense of having lost its immediate topical relevance, but also in the wider sense that we have been discussing. Shaw gave a more accurate picture of the nature of his play in an interview in 1904, in which he claimed the construction as a triumph: 'Just consider my subject—the destiny of nations! Consider my characters —personages who stalk on the stage incarnating millions of real, living, suffering men and women. Good heavens! I have had to get all England and Ireland into three hours and a quarter. I have shown the Englishman to the Irishman and the Irishman to the Englishman, the Protestant to the Catholic and the Catholic to the Protestant.'[18] The play dramatizes large historical developments by presenting the encounter between representative characters.

John Bull's Other Island introduces a historical dimension in an obvious way in the presence on the stage of the round tower as a conspicuous part of the set, which the characters speculate

---

[18] [Clement Shorter], 'George Bernard Shaw—A Conversation', *The Tatler*, 177 (16 Nov. 1904), 242.

about; Shaw elsewhere described such towers as 'inexplicable relics of a bygone social order'.[19] A more definite historical context is established for the play when the representative Englishman's valet, Hodson, who 'cant bear' the Irish, exclaims: 'Send em back to ell or C'naught, as good aowld English Cramwell said' (Act III, *CPP* ii. 947, 976). Tom Broadbent and his valet are the English invasion force of the play; Broadbent's method of subduing Ireland is much friendlier and more humane than Cromwell's was in the middle of the seventeenth century, but it is at least as effective. Instead of pulverizing Ireland with the power of an English army, he buys the country with the power of international Capitalism. (And at the same time he makes off with the local 'heiress' and in effect wins the Parliamentary seat, so that he will represent Rosscullen as a Liberal in Westminster.) Keegan summarizes the plot of the play in calling Broadbent '[t]he conquering Englishman' (Act IV, *CPP* ii. 1010).

The historical dimension of the play is fully established by the debate between Broadbent and Keegan in the last act. Broadbent and his colleague Doyle explain to Keegan that their syndicate already owns half of Rosscullen, and they will efficiently acquire the rest by lending mortgage money and then foreclosing, so that they can turn the area into a hotel and golf links. 'The world belongs to the efficient', Broadbent informs Keegan. The unfrocked priest responds with a long diatribe against the partners' business plans, concluding with a historical insight and a prophecy. 'For four wicked centuries', he proclaims, 'the world has dreamed this foolish dream of efficiency; and the end is not yet. But the end will come' (Act IV, *CPP* ii. 1017–18). This speech places *John Bull's Other Island* in the context of post-medieval European history, as the dramatization of a phase in the succession of epochs. The subject of the play is succinctly expressed in a phrase Shaw used in a Fabian essay written in the same year (1904): 'the anarchic period of transition from medievalism to modern collectivism'.[20] Like *Saint Joan*, Shaw's Irish play depicts the transition between one era and another; *Saint Joan* depicts the transition between the Catholic Middle Ages and the Protestant

---

[19] Shaw, *The Perfect Wagnerite, Shaw's Music*, iii. 445.
[20] Shaw, 'The Common Sense of Municipal Trading', *Essays in Fabian Socialism* (London: Constable, 1949), 205.

Renaissance, *John Bull's Other Island* the transition between the Capitalist era that Joan's Reformation and Renaissance brought into being, and a new era that will supplant it. Within *Saint Joan*, if we exclude the prophetic Epilogue, the medieval Church is triumphant and burns the proto-Protestant. Within *John Bull's Other Island*, if we exclude the implications of Keegan's prophecy that 'the end will come', Capitalism is triumphant and takes over the land of Ireland for its own unholy purposes. Each of these plays is set at the culmination of an era, when the predominant forces of the era retain their power, but when the prophetic observer can see that the end will come. The counterpart of Saint Joan as prophetic voice in *John Bull's Other Island* is Keegan, who has '*the face of a young saint*' and talks about Ireland as 'the island of the saints' (Acts II, IV, *CPP* ii. 922, 1016), who would like the English to leave his country, and who has been cast out by the Catholic Church. (And note that Keegan's reference to Ireland as 'holy ground' that Broadbent profanes (Act IV, *CPP* ii. 1016, 1019) has a counterpart in Joan's rebuke to Dunois: 'And you would stop while there are still Englishmen on this holy earth of dear France!' (Scene v, *CPP* vi. 146).) What Keegan prophesies, however, is the end of the epoch that Joan inaugurated, and what he represents is the noblest aspects of the medieval Catholicism that Joan helped to destroy.

'Four wicked centuries' would take one back to 1504, half a century after Joan's Rehabilitation—during the lifetime of Luther and just before the reign of Henry VIII in England. For the Catholic Keegan, the four wicked centuries are the centuries of Protestantism; his chronology links Capitalism with Protestantism. Keegan's reading of history is anti-Whig, and akin to that of Disraeli or G. K. Chesterton. Shaw wrote in 1933 that Chesterton's conversion to Roman Catholicism 'has obliged him to face the problem of social organization fundamentally, discarding the Protestant impostures on English history which inspired the vigorous Liberalism of his salad days' (Preface to *Too True To Be Good*, *CPP* vi. 412); and in 1917, when he reviewed Chesterton's *A Short History of England*, he related this book specifically to Keegan's speech:

Nearly fifteen years ago, in a play called John Bull's Other Island, I shewed an inspired (and consequently silenced) Irish priest saying to a couple of predatory commercial adventurers that 'for four wicked centuries the world has dreamed this foolish dream of efficiency; and the end is not yet. But the end will come.' If anyone had asked me then why I fixed that date (to do the British public and the critics justice, nobody ever did), I should not have been able to refer them to any popular history for an explanation. In future I shall be able to refer them to Mr Chesterton's. For Mr Chesterton knows his epochs, and can tell you when the temple became a den of thieves.[21]

This passage sounds decidedly pro-Chesterton and anti-Protestant, as does a 1911 reference to 'the extremely debased forms of religion which have masqueraded as Christianity in England during the period of petty commercialism from which we are emerging' ('A Foreword to the Popular Edition of "Man and Superman"', *CPP* ii. 532). We know, however, that Shaw's attitude towards this subject was not so simple. Like Max Weber and R. H. Tawney, he related Protestantism and the rise of Capitalism, but he both condemned and admired this historical development. Thus there is the same kind of conflict within *John Bull's Other Island* that we found in *Saint Joan*. On the one hand the play invites us to join Keegan in denouncing the four wicked centuries of Renaissance and post-Renaissance European history, but on the other hand the English Capitalist in the play is by no means an unattractive figure. His buoyancy and energy are appealing, and we are forced to concede that his hotel will be an improvement over the economic arrangements of the Irish people themselves. Shaw's play makes us take both sides at once: we agree with Keegan and Chesterton about the four wicked centuries, and we agree with Broadbent that the world (at the moment) belongs to the efficient. Broadbent's hotel is both a desecration of the island of the saints, and a necessary and beneficial economic enterprise.

In Victorian England, there was a buoyant, Whig member of Parliament, with an abundant faith in material progress, and a view of the Irish (in the early seventeenth century, at any rate) as 'distinguished by qualities which tend to make men interesting rather than prosperous'.[22] He was known to his contemporaries

---

[21] Shaw, 'Something Like a History of England', *Pen Portraits and Reviews* (London: Constable, 1949), 90–1.
[22] T. B. Macaulay, *The History of England from the Accession of James II*, *The Works of Lord Macaulay*, ed. Lady Trevelyan (8 vols., London: Longmans, Green, 1879), i. 51 (chap. 1).

as 'cocksure Tom' and described by one of them as 'an emphatic, hottish, really forcible person, but unhappily without divine idea'.[23] Shaw wrote in 1944 that 'Our Liberal friends of India, with their wits in Macaulay's history of the seventeenth century, sometimes talk as if our duty to India is to cease all persecutions and establish freedom of thought, speech, worship and education there',[24] which is a very Broadbentian stance (although Macaulay himself did not think of India in this way). Irish affairs occupy a significant part of Macaulay's *History of England*; and he depicts the English wars in Ireland (under William III, for example) as conflicts between more advanced and more primitive stages of civilization. His descriptions of English, Protestant victories over Irish Catholics set the stage, one might say, for Shaw's dramatic treatment of the subject in *John Bull's Other Island*.

I am not arguing that Shaw had Macaulay distinctly in mind when he created his quintessentially English Tom Broadbent—although he may well have done. What I do want to suggest is that *John Bull's Other Island* must be seen against the background of Victorian attitudes towards history. We are reminded of the Victorian intellectual background when Broadbent, near the end of the play, says of Keegan, 'What a regular old Church and State Tory he is! He's a character: he'll be an attraction here. Really almost equal to Ruskin and Carlyle' (Act IV, *CPP* ii. 1021). The debate that forms the culmination of the play is an encounter between two antithetical Victorian traditions.

One of the more notorious features of Macaulay's *History of England* is its habit of contrasting the primitive state of a place in the seventeenth century with the transformation that progress has effected by the nineteenth. The best-known example of this, I think, would be the description of Torbay then and now, in the narrative of William's landing there in November 1688.

Since William looked on that harbour its aspect has greatly changed. The amphitheatre which surrounds the spacious basin, now exhibits everywhere the signs of prosperity and civilisation. At the northeastern

---

[23] Hugh Trevor-Roper, Introduction to Macaulay's *The History of England* (Harmondsworth: Penguin, 1979), 27; J. A. Froude, *Thomas Carlyle: A History of the First Forty Years of His Life 1795–1835* (2 vols., London: Longmans, Green, 1882), ii. 231. Another Victorian Tom who may be relevant to Broadbent is Dickens's businessman in his condition-of-England novel *Hard Times*, Mr Gradgrind.

[24] Shaw, *Everybody's Political What's What?* (London: Constable, 1944), 151.

extremity has sprung up a great watering place, to which strangers are attracted from the most remote parts of our island by the Italian softness of the air. . . . The inhabitants are about ten thousand in number. The newly built churches and chapels, the baths and libraries, the hotels and public gardens, the infirmary and the museum, the white streets, rising terrace above terrace, the gay villas peeping from the midst of shrubberies and flower beds, present a spectacle widely different from any that in the seventeenth century England could show. . . . Torbay, when the Dutch fleet cast anchor there, was known only as a haven where ships sometimes took refuge from the tempests of the Atlantic. Its quiet shores were undisturbed by the bustle either of commerce or of pleasure; and the huts of ploughmen and fishermen were thinly scattered over what is now the site of crowded marts and of luxurious pavilions.[25]

This is exactly the transformation that Broadbent intends to bring about in Rosscullen, from primitive, inefficient farming to a Garden City that sounds strikingly similar to Macaulay's description of the transformed Torbay.

BROADBENT [*stopping to snuff up the hillside air*]. Ah! I like this spot. I like this view. This would be a jolly good place for a hotel and a golf links. Friday to Tuesday, railway ticket and hotel all inclusive. I tell you, Nora, I'm going to develop this place. . . .

BROADBENT. . . . I shall bring money here: I shall raise wages: I shall found public institutions: a library, a Polytechnic (undenominational, of course), a gymnasium, a cricket club, perhaps an art school. I shall make a Garden city of Rosscullen: the round tower shall be thoroughly repaired and restored.

(Act IV, *CPP* ii. 1006, 1015.)

Macaulay's paragraph could be taken as a gratuitous piece of Victorian Whig self-congratulation, but it is not really extraneous to the whole design of his *History*. His digression comes at the moment William landed on English soil, and the theme of the *History* is that all of this splendid achievement flows directly from William's successful invasion in 1688, with the resulting ascendancy of the values of Protestantism, commerce, and political stability. William's landing, in short, firmly established the era of which Broadbent is a part. While our feelings about Broadbent and his plans are highly ambivalent, there is an undeniable

[25] Macaulay, *History of England*, ii. 253–4 (chap. IX).

element of Macaulayite attraction towards material progress in Shaw's play.

Broadbent has the last word in *John Bull's Other Island*, and that word is 'hotel'. 'I feel sincerely obliged to Keegan', he tells Larry Doyle. '[H]e has made me feel a better man: distinctly better. [*With sincere elevation*] I feel now as I never did before that I am right in devoting my life to the cause of Ireland. Come along and help me to choose the site for the hotel' (Act IV, *CPP* ii. 1021–2). The hotel is the appropriate enterprise for Broadbent, in that it signifies bourgeois comfort and indicates the later stages of Capitalism. The hotel as an expression of bourgeois values is present too in a minor way in *The Man of Destiny*, in the contrast between the heroic Napoleon and the unheroic Italian innkeeper Giuseppe; and in *Arms and the Man* the fact that Bluntschli has inherited 'a lot of big hotels' from his father reinforces the contrast between the efficient bourgeois Swiss and the backward Bulgarians. A play in which hotels are linked with the theme of material progress is *The Millionairess*, in which Epifania transforms '*[a] dismal old coffee room in an ancient riverside inn*' into '*the lounge of . . . a very attractive riverside hotel*', anticipating the similar transformation in *Why She Would Not*. In both cases the renovation is an improvement. The stage directions in *The Millionairess* dwell on such details as the ugly Victorian wallpaper, a glass case containing an enormous stuffed fish, and an old-fashioned hat-stand which are replaced with '*[a]ll the appurtenances of a brand new first class hotel lounge*' (Acts II, IV, *CPP* vi. 917, 942).

Macaulay's description of Torbay is also brought to mind by another of Shaw's Garden Cities, Perivale St Andrews in *Major Barbara*. As with the new Rosscullen, its creator is a successful Capitalist, and as in *John Bull's Other Island*, many of the details take one back to Macaulay's Torbay. The company town that Andrew Undershaft has created '*is an almost smokeless town of white walls, roofs of narrow green slates or red tiles, tall trees, domes, campaniles, and slender chimney shafts, beautifully situated and beautiful in itself*'. In addition to the whiteness and the trees, there are counterparts of Torbay's infirmary and libraries:

SARAH. Heavens! what a place! . . . Did you see the nursing home!?
STEPHEN. Did you see the libraries and schools!?

SARAH. Did you see the ball room and the banqueting chamber in the Town Hall!?

(Act III, *CPP* iii. 157, 158–9.)

Major Barbara, like *John Bull's Other Island*, has its roots in English history. 'The Undershafts', Lady Britomart explains to her son Stephen in the first act, 'are descended from a foundling in the parish of St Andrew Undershaft in the city. That was long ago, in the reign of James the First' (*CPP* iii. 74). In this play the emphasis is more on Capitalism than Protestantism, and instead of the four wicked centuries of *John Bull's Other Island*, we have three centuries during which the Undershafts and their armaments foundry have represented the underlying reality of English society. The historical context of the play is also established by Undershaft's account of his six ancestors' mottoes over the centuries, with a comment that the fourth Undershaft 'did not write up anything; but he sold cannons to Napoleon under the nose of George the Third' (Act III, *CPP* iii. 168); and his opinion that 'history tells us of only two successful institutions: one the Undershaft firm, and the other the Roman Empire under the Antonines' (Act I, *CPP* iii. 75)—Gibbon's favourite historical period. While Undershaft represents the three Capitalist centuries, Cusins is the spokesman for the Hellenism of ancient Greece, and Peter Shirley for Victorian Secularism. Victorian history also enters the play in Lady Britomart's reference to Bismarck, Gladstone, and Disraeli (Act I, *CPP* iii. 72).

The pervasive Edwardian theme of inheritance is crucial in *Major Barbara*. The central issue of the plot is 'the Undershaft inheritance': who is to inherit the foundry—that is, who is to inherit the financial and military power of England? Inheritance enters the play, too, in the allusions to Undershaft's ancestry, Lady Britomart's aristocratic father, Cusins's parents in Australia, Snobby Price's Chartist parents who named him after the Chartist leader Bronterre O'Brien, and Rummy Mitchens's mother who named her Romola after the beautiful, spirited heroine of George Eliot's historical novel about late fifteenth-century Florence. (*Romola*, like *Major Barbara*, is a study of the religious temperament, in its treatment of Savonarola's fanaticism.) All of the characters have links with the past, and the audience is made to see the present as part of a historical continuum. And the debate between Undershaft and Cusins in Act III is really a discussion about the nature of the historical process. 'When you vote, you only change the names of the cabinet. When you shoot, you pull down governments, inaugurate new epochs, abolish old orders and set up new. Is that historically true, Mr Learned Man, or is it not?', Undershaft challenges the Professor of Greek; and 'Whatever can blow men up can blow society up. The history of the world is the history of those who had courage enough to embrace this truth' (Act III, *CPP* iii. 174–5). Cusins, the Hellenist, embraces this truth by the end of the play, and he intends to inaugurate a new epoch to supersede the aimless plutocracy of the three Undershaft centuries. *Major Barbara* is a thorough condition-of-England play, which dramatizes the signs of the times as part of the continuing process of history.

In *Heartbreak House*, the names of many of the characters place the play in a historical context. Mazzini Dunn, like Snobby (that is, 'Bronterre O'Brien') Price, was named by Victorian parents after a nineteenth-century man of action. Hesione tells Ariadne that Ellie's father 'is a very remarkable man. . . . His name is Mazzini Dunn. Mazzini was a celebrity of some kind who knew Ellie's grandparents. They were both poets, like the Brownings; and when her father came into the world Mazzini said "Another soldier born for freedom!" So they christened him Mazzini; and he has been fighting for freedom in his quiet way ever since. Thats why he is so poor' (Act I, *CPP* v. 71). In these cases of men who bear the names of a nineteenth-century Chartist leader or Italian patriot, their parents' choices turn out to be reminders of their inadequacy. Then there is the offstage character in *Heartbreak House* who has been governor of all the Crown colonies in succession, Sir Hastings Utterword, a name which suggests the Battle of Hastings and also Warren Hastings, the unscrupulous eighteenth-century Governor-General of India who is the subject of one of Macaulay's best essays.

Sir Hastings's wife, Ariadne, is named from Greek legend, and her sister and husband, Hesione and Hector, have names that are connected with the Trojan War: the original Hesione was the daughter of Laomedon, king of Troy, and Hector was the son of Priam, the leader of the Trojan forces during the siege. One of Hector's assumed names is Marcus Darnley, which takes us back to sixteenth-century Scottish history; Lord Darnley was the handsome, dissipated second husband of Mary Queen of Scots. But the most important historical allusions are the Trojan ones. Yeats, in a stanza of 'Two Songs from a Play', links the fall of Troy with the birth of Christianity as two moments of historical transformation, the frightening transition from one epoch to the next.

> Another Troy must rise and set,
> Another lineage feed the crow,
> Another Argo's painted prow
> Drive to a flashier bauble yet.
> The Roman Empire stood appalled:
> It dropped the reins of peace and war
> When that fierce virgin and her Star
> Out of the fabulous darkness called.[26]

These lines, like a number of Yeats's poems about the end of historical epochs ('The Second Coming', for example), give us a valuable context within which to consider *Heartbreak House*. The original Hector and Hesione were on the losing side of the Trojan War; they were part of the civilization that was doomed to disappear, to be supplanted by the Greeks. Shaw's play, like Homer's *Iliad*, is about a world-historical moment, and the moment in *Heartbreak House* is the ending of our civilization with the coming of the First World War. The use of the names Hesione and Hector in the play is another example of the way in which Shaw sweeps over history, finding analogous events in widely disparate ages. There is also an example of this historical levelling in the Preface to *Heartbreak House*, where we find another analogy between ancient Greek history and present history: 'Ever since Thucydides wrote his history, it has been on record that when the angel of death sounds his trumpet the pretences of civilization are blown from men's heads into the mud like hats in a gust of wind. But when this scripture was fulfilled among us, the shock was not the less appalling because a few students of Greek history were not surprised by it' (*CPP* v. 28).

Further historical allusions in *Heartbreak House* include Mangan as a Napoleon of industry, along with other references to

---

[26] W. B. Yeats, 'Two Songs from a Play', *The Collected Poems of W. B. Yeats* (London: Macmillan, 1961), 239–40.

Napoleon; Ellie's quotation from Walt Whitman's poem about Abraham Lincoln, 'O Captain, my captain' (Act III, *CPP* v. 176);[27] and Hector's judgement that the Heartbreakers 'are useless, dangerous, and ought to be abolished' (Act III, *CPP* v. 159), which Shaw told his French translator is 'No use in French. It is the wording of the resolution by which the Long Parliament abolished the House of Lords.'[28] And more than in any of Shaw's other plays, the plot of *Heartbreak House* looks to the past. Mangan's proposal to Ellie, Mazzini Dunn's business failures and partnership with Mangan, Billy Dunn's marriage to Nurse Guinness, and Shotover's marriage to the witch of Zanzibar all surface in the plot, and Shotover sees himself as merely a feeble echo of his own vigorous past. 'I can give you the memories of my ancient wisdom: mere scraps and leavings', he says to Ellie in Act II (*CPP* v. 146), and then in Act III he responds to Ellie's assertion that his spirit is not dead with 'Echoes: nothing but echoes. The last shot was fired years ago' (*CPP* v. 176). The name 'Shotover' itself suggests a man who is spent,[29] as do the names Dunn and Hushabye (or Hushabye at least may suggest someone who is temporarily sleeping).

These names, and the references to events of the past, are part of the whole pattern of meaning in *Heartbreak House*. Very few people, Shaw speculated in the Preface to the play, 'grasped the war and its political antecedents as a whole in the light of any

---

[27] When Shaw's Polish translator asked about this line of Ellie's, Shaw replied that 'It is an intentional quotation from Whitman. Ellie is a singer, and knows Cyril Scott's setting of the Lincoln poem' (Floryan Sobienowski to Shaw, 17 Mar. 1930, with Shaw's answer written on it in holograph, MS in Harry Ransom Humanities Research Center, the University of Texas at Austin).
[28] Shaw's revisions to A. and H. Hamon's French translation of *Heartbreak House*, MS in Harry Ransom Humanities Research Center, the University of Texas at Austin.
[29] One possible source for Shotover's name (apart from Carlyle's Captain of Industry, Plugson of Undershot in *Past and Present*) is in Shakespeare's *Henry V*, III. vii, in an exchange between Orleance and the Constable of France:

ORL. You are the better at proverbs, by how much 'A fool's bolt is soon shot.'
CON. You have shot over.
ORL. 'Tis not the first time you were overshot.

(ll. 121–4; *The Riverside Shakespeare*, ed. G. B. Evans *et al.* (Boston: Houghton Mifflin, 1974), 954.)
Here perhaps is another allusion to Shakespeare's history plays in Shaw's 'plays present'.

philosophy of history or knowledge of what war is' (*CPP* v. 35–6). The play gives only a slight idea of what war is, in its depiction of the Zeppelin raid in Act III, but it does try to enable us to grasp the First World War in the light of a philosophy of history. The inhabitants of the house, and indeed all of England and all of contemporary European civilization, are spent. The legacy of the past to the present is one of destruction; the present generation will pay not only for their own sins but for the sins of their nineteenth-century fathers as well. One of the revelations of the play's philosophy of history is that the War represents the end of our era; the War has the kind of significance that Yeats attached to the destruction of Troy or the birth of Christ. The heartbreak—or feeling of disillusionment—in the play presages the collapse of the civilization that the house signifies.[30] As Shaw wrote in a newspaper when the play opened in London in 1921, the heartbreak 'gets worse until the house breaks out through the windows, and becomes all England with all England's heart broken' ('Bernard Shaw on "Heartbreak House"', *CPP* v. 184).

The philosophy of history in the play also reveals that it is futile to rely on Providence to preserve one's civilization, and that such reliance is fatal in that it leads to inertia. In particular, the play constitutes a violent refutation of any Whig notion that England enjoys the special protection of Providence. The character who has come to repose his faith in Providence is the fatuous Mazzini Dunn, who sees the inhabitants of the house as 'rather a favorable specimen of what is best in our English culture . . . very charming people, most advanced, unprejudiced, frank, humane, unconventional, democratic, free-thinking, and everything that is delightful to thoughtful people' (Act III, *CPP* v. 173). Dunn would presumably have concurred in Macaulay's judgement on his own time as 'the most enlightened generation of the most enlightened people that ever existed', and with the grateful conclusion of Macaulay's Whig predecessor, the early nineteenth-century historian Henry Hallam, that there is a 'chain of causes through which a gracious providence has favoured the consolidation of our liberties and

---

[30] Hegel wrote of the Roman world: '[T]he world is sunk in melancholy: its heart is broken, and it is all over with the Natural side of Spirit, which has sunk into a feeling of unhappiness. Yet only from this feeling could arise the supersensuous, the free Spirit in Christianity' (G. W. F. Hegel, *The Philosophy of History*, tr. J. Sibree (New York: Dover, 1956), 278).

welfare'.[31] Dunn's view of history is that 'Nothing ever does happen. It's amazing how well we get along, all things considered.... Though I was brought up not to believe in anything, I often feel that there is a great deal to be said for the theory of an overruling Providence, after all' (Act III, *CPP* v. 175).

Against this Macaulayite optimism there is the philosophy of history expressed by Captain Shotover and really by the play itself, which is very much reminiscent of Carlyle. The last part of the play is another discussion of the nature of the historical process, with Shotover's fierce response to Dunn's confident assertion that '[n]othing happens'. Nothing happens at sea, he says, except 'the smash of the drunken skipper's ship on the rocks, the splintering of her rotten timbers, the tearing of her rusty plates, the drowning of the crew like rats in a trap' (Act III, *CPP* v. 176). He is apparently proved right by the bombing that follows, and whereas *John Bull's Other Island* seems to strike a balance between Macaulay and Carlyle, *Heartbreak House* leaves no room for Macaulay, and is informed by a view of history that is essentially Carlyle's.

A good way of exploring the Carlylean nature of *Heartbreak House*[32] is to consider the play's relationship to Carlyle's treatment of the French Revolution. Carlyle saw the French Revolution as an act of historical, divine retribution. The ruling classes in France had lost touch with the nature of things, and the nation must reap the terrifying harvest that grows from these failures. The harvest is one of the master metaphors of *The French Revolution*; others are the thunderstorm and the diabolical underworld, as in *Heartbreak House*. The Revolution is an appalling affliction on France, but it was deserved, and it has a purifying quality in that it is at least something *real* to replace the shams that dominated the country during the eighteenth century. Carlyle's descriptions of the French aristocracy in the eighteenth century would apply reasonably well to Shaw's Heartbreakers: 'It

---

[31] Macaulay, quoted in G. P. Gooch, *History and Historians in the Nineteenth Century* (1913; London: Longmans, 1952), 281; Hallam, quoted in J. W. Burrow, *A Liberal Descent*, 31.

[32] For discussions of some of the Carlylean elements in *Heartbreak House*, see Julian B. Kaye, *Bernard Shaw and the Nineteenth-Century Tradition* (Norman: Univ. of Oklahoma Press, 1958), 13–18; and J. L. Wisenthal, *The Marriage of Contraries: Bernard Shaw's Middle Plays* (Cambridge, Mass.: Harvard Univ. Press, 1974), 151–66.

is an unbelieving people; which has suppositions, hypotheses, and froth-systems of victorious Analysis; and for *belief* this mainly, that Pleasure is pleasant. Hunger they have for all sweet things; and the law of Hunger: but what other law? Within them, or over them, properly none!' The falseness and dishonesty of these people could not continue, for 'a Lie cannot endure for ever'. In the violence of the Revolution, the divine power of the universe asserted itself: 'Man's Existence had for long generations rested on mere formulas which were grown hollow by course of time; and it seemed as if no Reality any longer existed, but only Phantasms of realities, and God's Universe were the work of the Tailor and Upholsterer mainly, and men were buckram masks that went about becking and grimacing there,—on a sudden, the Earth yawns asunder, and amid Tartarean smoke, and glare of fierce brightness, rises SANSCULOTTISM, many-headed, fire-breathing, and asks: What think ye of *me*?'[33] Similarly the hollow formulas and the pleasure-seeking of the inhabitants of Heartbreak House are visited by a violent explosion that will blow their false, unbelieving, enervated world apart.

Carlyle sees the eighteenth century as Shaw sees the nineteenth: as a period of irreligion. In the first chapter of his *Frederick the Great* there is a sentence that seems particularly close to *Heartbreak House*, especially when we recollect the importance of the bankruptcy motif in the play, and also this dialogue between Hector and Ellie as the bombers come closer, and Hector has turned on all the lights in the house and torn down the curtains.

HECTOR. . . . There is not half light enough. We should be blazing to the skies.

ELLIE [*tense with excitement*]. Set fire to the house, Marcus.

(Act III, *CPP* v. 179.)

Here, from *Frederick the Great*, is Carlyle's comment on the French Revolution as the fitting termination of the eighteenth century:

[33] Carlyle, *The French Revolution*, i. 36–7, 212, *Works*, ii (Part I, Book i, chap. III; Part I, Book vi, chap. I). One of the many passages from Carlyle that suggest images and themes in *Heartbreak House* is in 'Corn-Law Rhymes', which asks about the state of England in 1832: 'Must it grow worse and worse, till the last brave heart is broken in England; and this same "brave Peasantry" has become a kennel of wild-howling ravenous Paupers?' (*Critical and Miscellaneous Essays*, iii. 159, *Works*, xxviii).

To resuscitate the Eighteenth Century, or call into men's view, beyond what is necessary, the poor and sordid personages and transactions of an epoch so related to us, can be no purpose of mine on this occasion. The Eighteenth Century, it is well known, does not figure to me as a lovely one; needing to be kept in mind, or spoken of unnecessarily. To me the Eighteenth Century has nothing grand in it, except that grand universal Suicide, named French Revolution, by which it terminated its otherwise most worthless existence with at least one worthy act;—setting fire to its old home and self; and going up in flames and volcanic explosions, in a truly memorable and important manner. A very fit termination, as I thankfully feel, for such a Century.[34]

Heartbreak House makes us value what is about to be destroyed more than this passage does, certainly, but in both Shaw's writing and Carlyle's there is the sense of history as inexorably just. According to Carlyle and Shaw the French Revolution and the First World War (respectively) could have been avoided, but they were appropriate punishments for falsehood and a failure to face reality. When Carlyle writes about his own time, he warns Victorian Englishmen to learn from what happened in France and to act accordingly to prevent a repetition in England. This is what Shaw is doing in *Heartbreak House*, but in writing the play during the War (in 1916–17) as he sees the historical pattern fulfilling itself, he not only warns about the future but points towards the punishment that is already being felt in the present.

[34] Carlyle, *History of Friedrich II of Prussia Called Frederick the Great*, i, Works, xii. 8–9 (Book i, chap. 1).

6

Shavian History and Shavian Drama

Shaw's claim in 1894 that 'you cannot even write a history without adapting the facts to the conditions of literary narrative' ('Ten Minutes with Mr Bernard Shaw', *CPP* i. 481) has an interesting bearing on the relationship between the work of the historian and that of the imaginative artist. We use the words 'history', 'story', and 'narrative' to refer to both types of work, and it is not easy to establish a clear boundary between them. Both the historian and the literary artist select from our experience in order to impose upon it a form that will convey meaning. In all historical writing—as opposed to chronicle—there is an element of the fictive, in that the historian shapes the historical record into a coherent narrative, whereas the chronicler merely lists events; and for the writers of historical classics—Clarendon or Gibbon, for instance—the patterning that gives aesthetic pleasure is a crucial aspect of their work.

It is in the case of the 'literary historian' that the dividing line between history and literature becomes especially difficult to draw, and the great age of literary historians was the nineteenth century. Two of the best examples of their art would be Carlyle's *French Revolution* and Macaulay's *History of England*. Both Macaulay and Carlyle learned from the historical fiction of Sir Walter Scott, and their histories are in many ways akin to the nineteenth-century novel. In his essay, 'History', however, Macaulay not only spoke about the need for the historian to reclaim those materials that the novelist has appropriated, but he commented, in his discussion of Tacitus, that 'The talent which is required to write history . . . bears a considerable affinity to the talent of a great dramatist.'[1] In his opening announcement of his intentions in *The History of England*, drama was the genre he drew on as a metaphor for the historical process. He explained his plan to give a 'slight sketch' of English history as background to

---

[1] T. B. Macaulay, 'History', *The Works of Lord Macaulay*, ed. Lady Trevelyan, (8 vols., London: Longmans, Green, 1879) v. 158, 144.

the detailed account of the period from 1685 until 1832: 'The events which I propose to relate form only a single act of a great and eventful drama extending through ages, and must be very imperfectly understood unless the plot of the preceding acts be well known.' One of the most gripping passages in the *History* is the trial in 1688 of the seven bishops of the Church of England who had challenged James II over the public reading of his Declaration of Indulgence. 'The trial then commenced,' Macaulay wrote, 'a trial which, even when coolly perused after the lapse of more than a century and a half, has all the interest of a drama.' Macaulay's account of the trial is shaped so that it will indeed have all the interest of a drama, or a novel. Earlier in the *History* he remarked of James's liaison with Catherine Sedley that '[A] dramatist would scarcely venture to bring on the stage a grave prince, in the decline of life, ready to sacrifice his crown in order to serve the interests of his religion, indefatigable in making proselytes, and yet deserting and insulting a virtuous wife who had youth and beauty for the sake of a profligate paramour who had neither.'[2] Macaulay saw historical events with the eye of the artist, and often with the eye of the dramatist (and he used Restoration plays as sources of evidence for the manners of that period).

The dramatic plot of Macaulay's *History* is that of comedy: William of Orange renews society by supplanting the blocking character, who as it happens is his father-in-law. The plot of Carlyle's *The French Revolution*, on the other hand, is tragic: the French nation suffers a horribly destructive but necessary upheaval as a result of a national flaw, and the work is written to inspire both pity and fear. The chapter on the conviction and execution of Robespierre in 1794 describes the events as the five acts of a '*natural* Greek Drama',[3] and Froude called the whole of *The French Revolution* 'an Aeschylean drama composed of facts literally true, in which the Furies are seen once more walking on this prosaic earth and shaking their serpent hair'.[4] In both

[2] Macaulay, *The History of England from the Accession of James II, Works*, i. 3 (chap. I); ii. 171 (chap. VIII); i. 578 (chap. VI).
[3] Thomas Carlyle, *The French Revolution*, iii. 281, *The Works of Thomas Carlyle*, ed. H. D. Traill (Centenary Edn., 30 vols., London: Chapman and Hall, 1896–9), iv (Part III, Book VI, chap. VII).
[4] J. A. Froude, *Thomas Carlyle: A History of His Life in London 1834–1881* (2 vols., London: Longmans, Green, 1884), i. 88.

Macaulay's history and Carlyle's we can see the historian as artist—sometimes as epic poet, sometimes as novelist, and sometimes as dramatist.

In most of Shaw's plays, conversely, we can see the dramatist as historian. It is not only his use of historical materials for a number of his plays, or his treatment of the present as a period of history. Shaw's conception of drama is closely related to the nature of the historian's work. In the first act of *'In Good King Charles's Golden Days'*, Isaac Newton is informed by his housekeeper that a 'Mr Rowley' and the Quaker George Fox have each called and wish to see him. Realizing that 'Mr Rowley' is in fact the King, Newton instructs the housekeeper to admit Fox if he calls again. 'Those two men ought to meet', he reflects (*CPP* vii. 218). Here, as Newton arranges the encounter of two people who 'ought to meet' because they represent diverse parts of their society and divergent ways of looking at the world, we can gain an insight into Shaw's concept of the nature of drama. One of Shaw's formulations on the subject was in a speech that he gave in 1928 to the Royal Academy of Dramatic Art:

If you look on that life as it presents itself to you, it is an extraordinarily unmeaning thing. It is just as if you took a movie camera and went out into the Strand or Piccadilly and began to turn the handle, and afterward developed your film and then said, 'Well, that is life—all those people moving about.' Lots of them have tragic histories, some of them have comic histories; some of them are abounding with joy because they are in love, others are going to commit suicide because they have been disappointed in love. It is all very wonderful! But when you look at the film you say, 'Well, I don't see anything there but a lot of people running about in a perfectly meaningless way.' Now what the drama can do, and what it actually does, is to take this unmeaning, haphazard show of life, that means nothing to you, and arrange it in an intelligible order, and arrange it in such a way as to make you think very much more deeply about it than you ever dreamed of thinking about actual incidents that come to your knowledge. That is drama, and that is a very important public service to render.[5]

That is drama, and one could say that it is history as well, in that the historian—unlike the chronicler—selects from the historical record and arranges his material in such a way as to make

---

[5] 'Bernard Shaw Talks about Actors and Acting', *Shaw on Theatre*, ed. E. J. West (New York: Hill and Wang, 1959), 198.

it intelligible and give it some kind of significance. In a comment in 1905 about *Mrs Warren's Profession*, Shaw denied that his business was to teach moral lessons. 'My business', he wrote, 'is to interpret life by taking events occurring at haphazard in daily experiences and sorting them out so as to show their real significance and interrelation' ('Shaw Replies to His Critics', *CPP* i. 362). This formulation, too, would apply to the work of the historian, as would a sentence in the Preface to *The Six of Calais*: 'Life as we see it is so haphazard that it is only by picking out its key situations and arranging them in their significant order (which is never how they actually occur) that it can be made intelligible' (*CPP* vi. 974). His chronicle plays, *Caesar and Cleopatra* and *Saint Joan*, can arrange situations in their significant order less than a history play like *Good King Charles*, and one of the reasons for the Epilogue in *Saint Joan* is to enable the dramatist to go beyond the chronicle form and to derive an order from the historical data. In his 'present history' plays Shaw is able to make significant connections without constraint, to bring together men and women who ought to meet—the English Capitalist and the Irish Medievalist, the arms manufacturer and the Professor of Greek, the colonial governor's wife and the burglar.

Shaw began his literary career as a novelist, and it was after he had written five novels and part of a sixth that he turned to the theatre in 1884, when he and William Archer collaborated in what was to become Shaw's first play in 1892. There are many explanations for Shaw's switch from novelist to dramatist, and some of them are related to his attitudes towards history. One could say that his particular sense of history led him from the novel into the drama, or that his particular sense of the world underlies both his historical outlook and his attraction to drama. But there is a connection between drama as a genre and Shaw's way of looking at history. The very fact that full-length plays are usually divided into acts—in spite of Shaw's experiments in *Getting Married* and *Misalliance*—brings drama closer to a cataclysmic as opposed to a gradualist view of history. Shaw, with his sense of epochs, could not have written a poem like Tennyson's *In Memoriam*, even though its meliorism anticipates one side of his own thinking. The form of *In Memoriam* is evolutionary and gradualist; there are no sharp breaks between one stanza and another, or between one section and another. Shaw, who found progression by semitones too gradual for his ardent nature, was attracted to sharp breaks. His expository prose, as in his prefaces, is divided into sections with separate headings, and his argument will often shift direction suddenly from one paragraph to another. There is a useful comparison to be made, I think, between Shaw's prose and some of Yeats's poems, such as 'The Tower' and 'Meditations in Time of Civil War', with their well-defined subsections, and also Yeats's superb prose with subsections that make an apparently new start. And Yeats, like Shaw, was a dramatist (not only in his plays but also in his poems), and like Shaw he saw history as a succession of distinct epochs.

In a play each act can begin a kind of new epoch, and history can be rendered as a series of sudden transformations, whereas the novel form is better suited to the presentation of subtle, gradual nuances of historical change. As Georg Lukács has suggested in his chapter on 'Historical Novel and Historical Drama' in *The Historical Novel*, the 'world-historical individual' is an appropriate hero in drama, whereas the novel (his main example being Scott) is properly concerned with the background to historical change rather than with the major encounters themselves.

The 'world-historical individual' can only figure as a minor character in the novel because of the complexity and intricacy of the whole social-historical process. The proper hero here is life itself; the retrogressive motifs, which express necessary tendencies of development, have as their hidden nucleus the general driving forces of history. The historical greatness of such characters is expressed in their complex interaction, their manifold connection with the diverse private destinies of social life, in whose totality the trends of popular destiny are revealed. In drama these historical forces are represented directly through the protagonists. Since the hero of drama unites in his personality the essential social-moral determinants of the forces which produce the collision, he is necessarily ... a 'world-historical individual'. Drama paints the great historical explosions and eruptions of the historical process. Its hero represents the shining peak of these great crises. The novel portrays more what happens before and after these crises, showing the broad interaction between popular basis and visible peak.[6]

[6] Georg Lukács, *The Historical Novel*, tr. H. and S. Mitchell (London: Merlin, 1962), 149–50.

If Lukács is right in this distinction between the historical novel and historical drama, then Shaw's transition from novelist to dramatist is an expression of his attitude towards the individual in history, the role of the hero. Drama gives Shaw the opportunity to embody historical forces in protagonists like Caesar and Joan, who are indisputably world-historical individuals; and it gives him the opportunity to express 'the shining peak' of great historical crises. And his world-historical individuals, I would argue, need not be literally historical, but can be fictional, symbolic embodiments of historical forces manifesting themselves in the individual will. One good example of a play that puts such characters on the stage would be *Major Barbara*. This play depicts the great explosions and eruptions of the historical process, expressing historical forces directly through the protagonists, even though this would not have been the sort of play that Lukács had in mind in that it is not in the narrow sense a history play at all. But it does, as we have seen, dramatize historical forces, and its dialectical conflict suggests a view of history according to which (in the words of *The Perfect Wagnerite*) 'human enlightenment does not progress by nicer and nicer adjustments, but by violent corrective reactions'.[7] Another good example of a play that Lukács's passage describes, even though it too is not at all the sort of thing he had in mind, would be *Back to Methuselah*, which expresses the radical transformations of history in its form: it consists of a series of separate plays, rather in the way that Yeats's 'Meditations in Time of Civil War', for example, is made up of a series of separate poems, or Wagner's music-drama of historical epochs, *The Ring of the Nibelung*, is made up of a series of separate operas.

The form of *Back to Methuselah* embodies Conrad Barnabas's insight into the way in which biological history proceeds. He and his brother are explaining the sudden emergence of Creative Evolution as the modern religion when the politician Lubin offers one of his conventional observations:

LUBIN. But surely any change would be so extremely gradual that——
CONRAD. Dont deceive yourself. It's only the politicians who improve the world so gradually that nobody can see the improvement. The

---

[7] *Shaw's Music*, ed. Dan H. Laurence (3 vols., London: Max Reinhardt, The Bodley Head, 1981), iii. 475.

notion that Nature does not proceed by jumps is only one of the budget of plausible lies that we call classical education. Nature always proceeds by jumps. She may spend twenty thousand years making up her mind to jump; but when she makes it up at last, the jump is big enough to take us into a new age.

(*The Gospel of the Brothers Barnabas, CPP* v. 429.)

The next play in the 'metabiological pentateuch' is *The Thing Happens*, in which the new epoch begins.

This title, *The Thing Happens*, points to a further connection between Shaw's dramatic form and his sense of history. Historical development in *Back to Methuselah* is not predestined; the thing happens because people will it to happen when the need becomes pressing. Shaw's world is not one of accident or blind chance on the one hand, nor of a fixed plan or immutable structure on the other. You never can tell when or how things will happen. When Shaw was asked by an interviewer whether the United States was gradually annexing the English-speaking world, along the lines suggested in *The Apple Cart*, he replied that this 'is still a matter for speculation. It is not impossible. But, so far, it cannot be said that the bond of Western European civilization is weaker than the bond of language. And do not forget that the Marxian dream of a world-wide proletarian revolution, though it is not now practical politics, may yet upset all our conceptions of international relations. The Reformation did not seem practicable in the Middle Ages; but it happened for all that' ('Mr Shaw and Democracy', *CPP* vi. 388–9).[8] You never can tell. Another of Shaw's titles which conveys this sense of human affairs is *The Simpleton of the Unexpected Isles*. The favourite phrase of one of the characters in this play is 'Let life come to you' (Prologue, *CPP* vi. 769), and the priestess and priest say near the end:

[8] Cf. 'The greatest political fact of your lifetime is that nothing has happened in the twentieth century except the impossible. The conversion of Tsarist Russia into a Communist Republic was wildly impossible. The swatting of the Hohenzollern and Holy Roman Empires like two torpid flies was ridiculously impossible. The two Labor Cabinets were impossible. The metamorphosis of the Intransigent Macdonald [*sic*] of the eighties into a Conservative leader and a shamelessly reactionary oratorical bunk merchant was impossible. The acceptance of *your* estimate of the British House of Commons by Lenin, by Mussolini, by Primo de Rivera, Horthy *et hoc genus omne*, was impossible. As Inge puts it, nothing has failed like success. As you must put it, nothing has succeeded like impossibility' (Shaw to St John Ervine, 28 Jan. 1932, MS in Harry Ransom Humanities Research Center, the University of Texas at Austin).

PROLA. ... Remember: we are in the Unexpected Isles; and in the Unexpected Isles all plans fail. So much the better: plans are only jigsaw puzzles: one gets tired of them long before one can piece them together. There are still a million lives beyond all the Utopias and the Millenniums and the rest of the jigsaw puzzles: I am a woman and I know it. Let men despair and become cynics and pessimists because in the Unexpected Isles all their little plans fail: women will never let go their hold on life. We are not here to fulfil prophecies and fit ourselves into puzzles, but to wrestle with life as it comes. And it never comes as we expect it to come.

PRA. It comes like a thief in the night.

PROLA. Or like a lover. Never will Prola go back to the Country of the Expected.

PRA. There is no Country of the Expected. The Unexpected Isles are the whole world.

(Act II, *CPP* vi. 839.)

The idea that history is alive and unexpected rather than predestined or accidental has its counterpart in Shaw's dramatic form in his rejection of the well-made play. Thus his sense of history is expressed in his choice of drama as a genre, and then in his choice of the kinds of play that he wrote (and did not write). In the late nineteenth century the novel would have been a more obvious choice for an aspiring writer than the drama, and within the theatre the well-made play would have been an obvious choice, for it was the established dramatic convention of the period. Under the influence of the French drama, particularly the plays of Eugène Scribe and Victorien Sardou, English playwrights favoured tightly constructed plots, in which accident and misunderstanding—as opposed to will—lead to a contrived outcome. Shaw said that when he began his dramatic collaboration with William Archer in the 1880s Archer looked for the conventional construction whereas he himself 'classed constructed plays with artificial flowers, clockwork mice, and the like' (cf. Prola's jigsaw puzzles). Archer 'did not agree with me that the form of drama which had been perfected in the middle of the nineteenth century in the French theatre was essentially mechanistic and therefore incapable of producing vital drama.... I held... that a play is a vital growth and not a mechanical construction.'[9]

---

[9] Shaw, 'How William Archer Impressed Bernard Shaw', *Pen Portraits and Reviews* (London: Constable, 1949), 7, 22.

Looking back in 1946 on his early years as a playwright, Shaw rejected the suggestion that he was of the school of Arthur Wing Pinero, Henry Arthur Jones, R. C. Carton, Sydney Grundy, and Oscar Wilde. 'I was furiously opposed to their method and principles', he wrote. 'They were all for "constructed" plays, the technique of construction being that made fashionable by Scribe in Paris. . . . Plays manufactured on this plan, and called "well-made plays," I compared derisively to cats'-cradles, clockwork mice, mechanical rabbits, and the like.'[10] His idea of a play is like his idea of history; a play is something living and growing, not proceeding according to a fixed plan, and the spring of action is the human will rather than trivial accident or any inevitable consequences of the past.

There is yet another correspondence between Shaw's sense of history and the dramatic form of his plays. In *The Quintessence of Ibsenism* he remarked that 'a difference of opinion between husband and wife as to living in town or country might be the beginning of an appalling tragedy or a capital comedy',[11] which implies that a playwright can transform the data of experience into a variety of dramatic expressions. Hayden White, in his rhetoric of nineteenth-century historiography, has argued that the historian is not compelled by the historical record itself to 'emplot' his story in any particular way, and White's exposition seems to me to have a distinct bearing on Shaw's practice as a dramatist. In his *Metahistory: The Historical Imagination in Nineteenth-Century Europe* (1973), White constructs elaborate schemata to analyse the historical writing of Hegel, Michelet, Ranke, Tocqueville, Burckhardt, Marx, Nietzsche, and Croce. What is particularly relevant to drama is his theory of emplotment. There is no such thing as 'realistic' historical writing that simply tells what really happened. The historian chooses among various possible modes of emplotment to tell his story, to render a sequence of events as a story, as history. As White explains his theory in a later essay, 'Considered as potential elements of story,

---

[10] Shaw, 'My Way with a Play', in *Shaw on Theatre*, 268. Martin Meisel, in *Shaw and the Nineteenth-Century Theater* (Princeton: Princeton Univ. Press, 1963), 353–5, discusses Scribe's *Le Verre d'eau* as an example of the nineteenth-century history play. This work, set in the court of Queen Anne, suggests that great historical effects are produced by trivial personal causes.

[11] Shaw, *The Quintessence of Ibsenism*, in *Shaw and Ibsen*, ed. J. L. Wisenthal (Toronto: Univ. of Toronto Press, 1979), 214.

historical events are value-neutral. Whether they find their place finally in a story that is tragic, comic, romantic or ironic—to use Frye's categories—depends upon the historian's decision to configure them according to the imperatives of one plot-structure or mythos rather than another.'[12]

In *Metahistory*, White emphasizes the internal tensions that characterize great historiography. He talks about inconsistencies between a historian's mode of emplotment and what he calls the 'mode of argument' and the 'mode of ideological implication'. For purposes of illuminating Shaw's dramatic practice, there is no need to pursue all of White's categories, but we might concentrate on the idea of dialectical tensions within the mode of emplotment. White applies the terminology of Northrop Frye's theory of myths to historical writing: 'Tragedy and Satire are modes of emplotment which are consonant with the interest of those historians who perceive behind or within the welter of events contained in the chronicle an ongoing structure of relationships or an eternal return of the Same in the Different. Romance and Comedy stress the emergence of new forces or conditions out of processes that appear at first glance either to be changeless in their essence or to be changing only in their phenomenal forms.' He argues that Hegel and Marx emplot history as *both* tragedy and comedy. Hegel's purpose, he summarizes, 'is to justify the transition from *the comprehension of the Tragic nature of every specific civilization* to *the Comic apprehension of the unfolding drama of the whole of history*'; while Marx's historical vision, similarly, 'oscillated between apprehensions of the Tragic outcome of every act of the historical drama and comprehensions of the Comic outcome of the process as a whole'.[13]

It is this encounter between genres that one finds in Shaw's dramatizations of the historical process. In most of Shaw's plays the dominant mode of emplotment is comic, but this mode does not dominate to the extent that it does in (let us say) Macaulay's

---

[12] Hayden White, 'The Historical Text as Literary Artifact', in Robert H. Canary and Henry Kozicki, eds., *The Writing of History: Literary Form and Historical Understanding* (Madison: Univ. of Wisconsin Press, 1978), 47–8, 55. See also *Metahistory: The Historical Imagination in Nineteenth-Century Europe* (Baltimore: The Johns Hopkins Univ. Press, 1973), 5–11.

[13] White, *Metahistory*, 29–31, 11, 117, 328. For further comments on Marx's emplotment of the historical process 'in two modes, Tragic and Comic, simultaneously', see 286–7, 310.

History. There is a strong pull in the contrary direction. *Caesar and Cleopatra*, for example, tells the story of the Great Man's success in Alexandria: the more progressive historical force is supplanting the more primitive. Much of the dramatic interest, however, centres on what Caesar *fails* to do. Caesar may conquer as a soldier but he gets almost nowhere as a teacher, and the more significant arena in the play is the mind of Cleopatra rather than the streets of Alexandria. In the end the victor leaves the world much as he found it, and the play compels the audience to adopt a perspective that encompasses the subsequent 2,000 years in which nothing has really improved. In *Major Barbara*, in which we look backwards from the present instead of forwards to it, once again history is emplotted as both comedy and tragedy. Here the play ends with the distinct possibility of comic renewal, but this possibility does not eradicate from our minds the sordid contemporary reality of the Salvation Army shelter in Act II, or the fact that poverty and war have existed all through the three Undershaft centuries of modern history and it is by no means inevitable that they will disappear now. *Man and Superman* also ends with the traditional renewal pattern of comedy, but in this play the Devil's ironic, tragic theory of cyclical recurrence stands in sobering counterpoint to the buoyant intimations of progress in the play. In *John Bull's Other Island* we have a comic emplotment of history in the sense that the Macaulayite apostle of progress is victorious over the primitive Irish, but he too is out of date—he embodies the end of the present epoch rather than the beginning of a new one. We are left with the ineffectual unfrocked priest's prophetic assurance that the end will come, but with the forces of the present epoch very much in control. Furthermore, the defeated, superseded Irish are in some measure protagonists in the play, sympathetically presented.

In *Heartbreak House* the present epoch *is* being swept (or bombed) away, but the emplotment of history here is mainly tragic. We are made to look at the historical process largely from the point of view of the victims, as opposed to whatever might supplant the disappearing civilization. The play does convey a slight suggestion of a clearing-away of debris (as Carlyle's *French Revolution* does), but the dominant tone is one that expresses tragic destruction. The civilization that is being swept away may be effete and irresponsible, but it is our own.

Back to Methuselah offers an especially good example of this mixed emplotment of the historical process, in that one of its sections is explicitly labelled as tragedy: *Tragedy of an Elderly Gentleman*. The Elderly Gentleman's story is tragic in that he is one of the last survivors of his era; the term 'tragedy' in the title indicates a particular phase of historical development, which may be emplotted as tragedy when seen from the perspective of the victim. But *Back to Methuselah*, in spite of this tragedy and in spite of the fall in the first two plays of the cycle, is for the most part a comedy, as we discover by the time we have achieved the perspective that is possible when we have gone as far as thought can reach. The cycle as a whole offers a much wider perspective than *Heartbreak House* does, and we are able to see the tragic demise of our epoch and ourselves in the larger context of the comedy of Life's progress. That is, the tragic nature of our specific civilization is encompassed by the comic apprehension of the unfolding drama of the whole of history—or as much of history as present thought can reach, since Shaw's historical vision, unlike Hegel's, draws attention to the unpredictable future.

In the Preface to *Saint Joan* there is a sentence which gives an important clue to the structure of that play: 'The romance of her rise, the tragedy of her execution, and the comedy of the attempts of posterity to make amends for that execution, belong to my play and not to my preface, which must be confined to a sober essay on the facts' (*CPP* vi. 66). The romance of the first three scenes of *Saint Joan* gives way to the tragedy of the next three, as the play goes on to show the struggle and defeat that result from Joan's rapid rise. It is true that the historical record demands that her story be taken further than her successes at Orléans and Rheims, but the point is that in *Saint Joan* the historical record and Shaw's genius find a remarkable meeting-ground (in this respect and in others too). The historical record, however, does not require the play's next—and more jarring—reversal. The romance of her rise is not enough for Shaw, and neither is the tragedy of her execution. Our attention is specifically drawn to an apparently tragic emplotment of Joan's history by the Archbishop's line in the Cathedral scene, 'The old Greek tragedy is rising among us. It is the chastisement of hubris' (*CPP* vi. 146). But then, after Joan's execution, the scene changes unexpectedly from the formal trial setting of the hall of the castle to the informal atmosphere of

Charles's bedroom, with the King '*reading in bed, or rather looking at the pictures in Fouquet's Boccaccio with his knees doubled up to make a reading desk*' (*CPP* vi. 190), and we discover we have entered the world of comedy in that Joan is now triumphant. The comedy is qualified by the new rejection of Joan that follows, and in the end one cannot label the play as a whole as romance, tragedy, or comedy.[14] The plot of history is one of tragic recurrence, and also one of possible comic linear ascent. Shaw has emplotted the historical record as a mixture of different modes, just as he sees history from a number of different perspectives. The plot of history, in Shaw's plays, is neither pure romance nor pure tragedy nor pure comedy, but a vital, unexpected encounter between antithetical ways of interpreting the historical process.

Saint Joan is the play that best represents the complexity and many-sidedness of Shaw's historical attitudes. Here we clearly see the mixed response to the military and spiritual hero, the attachment to both the Middle Ages and the Renaissance, the interplay between past and present, and the antithetical views of progress that (as we have just been noting) are reflected in competing plot structures. *Saint Joan* demonstrates the way in which Shaw's dramatic treatment of historical patterns does not take an audience in one direction only. There is a tension between opposing ways of understanding history, a tension that compels an audience to join with the play in exploring the dialectical conflicts. Our concern here is not with Shaw's stature as a historian *per se*, or as a philosopher of history, but rather with Shaw as a dramatist whose plays are enriched by their simultaneous perception of opposing ways of approaching the process of history.

*

I hope this study has demonstrated something of the vitality and complexity of Shaw's mind and work. 'The man who never alters his opinion is like standing water, & breeds reptiles of the mind,' wrote Blake in *The Marriage of Heaven and Hell*.[15] It is not so

---

[14] Part of this discussion of *Saint Joan* is taken from my article, 'Having the Last Word: Plot and Counterplot in Bernard Shaw', *ELH* 50 (Spring 1983), 175–96.

[15] William Blake, *The Marriage of Heaven and Hell*, *The Complete Writings of William Blake*, ed. Geoffrey Keynes (London: Oxford Univ. Press, 1966), 156 (Plates 17–20).

much that Shaw changed his mind—although he certainly did so on questions like the military Great Man—as that his mind was large and active enough to encompass more than one point of view at a time. He described himself in the Preface to *Three Plays for Puritans* as a crow who had followed many ploughs (*CPP* ii. 47), and his eclecticism has been perceptively described by Edmund Wilson: 'So far from being relentlessly didactic, Shaw's mind has reflected in all its complexity the intellectual life of his time.'[16] In examining Shaw's sense of history, we have seen him as a crow following many Victorian ploughs, and we have seen plenty of evidence to support Wilson's statement, and also Jacques Barzun's apt characterization of Shaw as 'a live repository of all the great thought and art of his century'.[17]

On almost all of the issues that we have been examining, Shaw's writing has taken both sides at once. He is both sympathetic and hostile to the Middle Ages, and he is both sympathetic and hostile to the Renaissance. He decisively rejects the idea of progress, and he enthusiastically adopts it. He is close to Carlyle, and he is also close to Macaulay. It is not that he is necessarily influenced by either of them, or by any particular Victorian figure, but that he comes out of the entire Victorian world of ideas. He is saturated with the attitudes of writers like Carlyle and Macaulay, and he is part of both the Utilitarian, Whig side of Victorian thinking and the anti-Utilitarian, Tory–Romantic side. Raymond Williams indicates this combination in his *Culture and Society*:

Shaw's association with Fabianism is of great importance, for it marks the confluence of two traditions which had been formerly separate and even opposed. Fabianism, in the orthodox person of Sidney Webb, is the direct inheritor of the spirit of John Stuart Mill; that is to say, of a utilitarianism refined by experience of a new situation in history. Shaw, on the other hand, is the direct successor of the spirit of Carlyle and of Ruskin, but he did not go the way of his elder successor, William Morris. In attaching himself to Fabianism, Shaw was, in effect, telling Carlyle and Ruskin to go to school with Bentham, telling Arnold to get together with Mill.[18]

[16] Edmund Wilson, 'Bernard Shaw at Eighty', *The Triple Thinkers* (1938; New York: Oxford Univ. Press, 1963), 184.
[17] Jacques Barzun, 'From Shaw to Rousseau', *The Energies of Art* (New York: Harper, 1956), 247.
[18] Raymond Williams, *Culture and Society 1780–1950* (London: Chatto and Windus, 1958), 181–2.

In his attitude towards progress, for example, Shaw combined the Whig–Utilitarian belief in improvement with the Tory–Romantic view of the present age (the nineteenth century) as one of particular horror and degradation. One could say that the Tory–Romantic standard is the past; the Whig–Utilitarian standard is the present; and Shaw's standard is the future, and that he joins the former in condemning the present, and the latter in condemning the past.

One might react to this eclecticism by protesting that Shaw's views are a muddle of contradictory assertions, and one could argue that Shaw never breeds reptiles of the mind because he is never able to make up his mind. I would respond to such an argument by asserting that Shaw's eclecticism is appropriate to the dramatist, that the real criterion is the quality of the plays, and that the intellectual and dramatic vitality of his plays arises in part from the encounter within them between antithetical Victorian intellectual traditions. The fact that he takes both sides at once is not a source of confusion in his work, but rather a source of dramatic energy. 'We make out of the quarrel with others, rhetoric, but of the quarrel with ourselves, poetry', wrote Yeats,[19] who regarded Shaw as all rhetorician and no poet at all. But from Shaw's quarrel with himself about the nature and meaning of history, there came great drama.

[19] W. B. Yeats, 'Per Amica Silentia Lunae', *Mythologies* (London: Macmillan, 1959), 331.

Index

Adams, Elsie B. 79 n.
Aknaton 109
Amanullah Khan 109
Ammianus Marcellinus 41
anachronism 101–8
Androcles and the Lion 25, 66
   anachronisms in 105, 107
   and historical fact 43–4, 50–1
   persecution in 110–11, 118
Anne, Queen 31
Appian of Alexandria 41
Apple Cart, The 29, 68, 69, 95, 143, 171
   as present history 149–50
Archer, William 168, 172
Aristophanes 36
Arms and the Man 65
   and historical fact 49
   hotels in 156
Arnold, Matthew 5, 18, 85, 142, 145, 147–8, 178
Atatürk *see* Mustafa Kemal Atatürk
Attila 138

Back to Methuselah 29–30, 31, 33, 37, 97–8, 100, 135–6, 137–8, 170–1
   as comedy and tragedy 176
   Great Man in 64–5
   as present history 147–8
   progress in 114–15, 117, 120–1, 122, 124–6, 133
Bacon, Francis 4–5
Barbier, Paul Jules 47 n.
Barzun, Jacques 178
Bax, Ernest Belfort 20
Beethoven, Ludwig van 36
Belisarius 27
Bellini, Vincenzo 17
Belloc, Hilaire 48
Bentham, Jeremy 178
Bismarck, Otto von 157
Blake, William 56, 177
Booth, Edwin 59
Bosworth Field, Battle of 15, 32
Boucicault, Dion 17
Boyne, Battle of the 96
Bright, John 99
Brown, Ford Madox 86 n.

Browning, Robert 23
Buckle, Henry Thomas 6–7, 23, 38 n., 51, 87, 95 n., 127
   on nineteenth century 9–10
   on progress 6–7
   Shaw on 20–22, 51–2
Buckley, Jerome Hamilton 13
Buddha 36
Bulwer-Lytton, Edward 16
Bunyan, John 17–18, 36, 58 n., 91, 129
   Shaw's knowledge of 16, 19
Buoyant Billions 34, 124
Burke, Edmund 31, 130
Burtsev, Vladimir 109
Butterfield, Herbert 105
Byron, George Gordon, Lord 36, 58 n.

Cade, Jack 27
Caesar and Cleopatra 28, 29, 33, 92, 132, 135–6, 168
   anachronisms in 101–3, 106–7
   as comedy and tragedy 175
   Great Man in 57, 58–62, 63, 69–70, 170
   and historical fact 24, 41–3, 50, 52–3, 55
   progress in 112–13, 119–20, 129–30
Candida 79, 85, 146
Capitalism 22, 31, 78, 85, 89–90, 91, 97, 98–9, 136–7, 150, 151–3, 156, 157
Carlyle, Thomas 1–12, 30, 117 n., 125, 148, 154
   on Cromwell 3–4, 11, 56, 57, 70, 71–2
   on George Fox 95–6
   on Frederick the Great 2, 3, 4, 57
   on French Revolution 11, 162–4
   French Revolution, The 2, 40, 165, 166–7, 175
   Great Man in 3–4, 56–8, 62
   and Great Man in Shaw's plays 68–75
   and *Heartbreak House* 160 n., 162–4
   on historical writing 103–4, 106
   on Luther 11, 56, 70, 74, 87
   on Macaulay 1–2
   on Mahomet 3, 25–6, 56, 70, 75

Index

Carlyle, Thomas – *cont.*
  on Middle Ages 2, 10, 11, 80, 82
  on nineteenth century 9–10, 98–9
  Oliver Cromwell's Letters and Speeches 2, 11
  and *On the Rocks* 149
  and present history 141–2
  on progress 7–9
  on Protestantism 2
  and *Saint Joan* 71–5
  Shaw and 22–3, 36–7, 58 n., 65–6, 97, 125, 178
Carr, E. H. 4, 40 n.
Carton, R. C. 173
Catholicism 82, 85, 89, 90, 94–5, 120, 151–2
Chamberlain, Houston Stewart 21
Charles I: 92, 109, 110, 149
Charles II: 92
  see also 'In Good King Charles's Golden Days'
Charles XII (Sweden) 64
Chekhov, Anton 6
Chesterton, G. K. 33, 77, 98, 152–3
Christianity 59 n., 118, 120, 159, 161
  see also Catholicism, Protestantism, Reformation
Clarendon, Earl of (1609–74) 165
Cockerell, Sydney 44
Collingwood, R. G. 5, 38, 132–3
Confucius 36
Constantine 136
Corday, Charlotte 29
Couchman, G. 89 n.
Creighton, Mandell 144 n.
Cromwell, Oliver 97, 110, 112, 149, 151
  Shaw on 26, 67–8, 69, 91–2
  Shaw's unwritten play on 25, 57, 58, 63
  see also Carlyle, Thomas
Crusades 70, 71
Culler, A. Dwight 13 n., 139 n.
cyclical view of history 116–18, 126, 132, 175

Danton, Georges Jacques 73
Dark Lady of the Sonnets, The 25, 54, 57
Darnley, Henry Stuart, Lord 159
Darwin, Charles 8, 99, 100, 132–3
Devil's Disciple, The 28, 118
  and historical fact 24, 41, 44, 49–50, 53, 54

Dickens, Charles 22, 57, 97, 98–9, 118, 154 n.
  Shaw's knowledge of 15
Diderot, Denis 36
Diodorus Siculus 41
Disraeli, Benjamin 43, 101–2, 141, 152, 157
Doctor's Dilemma, The 58
Dover, Treaty of 93–5
Drinkwater, John 25 n.
Dumas, Alexandre (Dumas *père*)
  Shaw's knowledge of 15

Eliot, George 157
Elizabeth I: 98
Emmet, Robert 27
Engels, Friedrich 32, 36
English Historical Review 105 n., 144 n.
epochs 19, 77–8, 108, 151–2, 159, 168–9
Erasmus, Desiderius 36
Ervine, St John 16 n.
Essex, Earl of (1566?–1601) 27
evolution 132–4, 170–1

Fanny's First Play
  as present history 144, 146–7
Farfetched Fables 34
  progress in 122
Fawkes, Guy 149
Feudalism 32, 82, 83, 85, 90, 91
First World War 27, 35, 63, 87–8, 100, 121, 125, 147–8, 159, 160–1, 164
Flinders Petrie *see* Petrie, William Matthew Flinders
Fonblanque, Edward Barrington de 24
Forbes-Robertson, Johnston 26 n.
Ford, Henry 27
Forster, E. M. 142, 144, 145
Fourier, François 36
Fowler, William Warde 41, 42
Fox, George 70, 112
  in 'In Good King Charles's Golden Days' 69, 95–6
  see also Carlyle, Thomas; Macaulay, Thomas Babington
France, Anatole 18, 47 n.
Franco, Francisco 149
Frederick the Great 4, 146
  see also Carlyle, Thomas
Freeman, Edward 139
French Revolution 28, 36, 66, 148
  see also Carlyle, Thomas

Index

Froissart, Jean 25, 53
Froude, James Anthony 2, 59 n., 166
Frye, Northrop 174

Galileo 89–90
Geneva 28, 88
  as present history 147, 149
George III: 157
Getting Married 70, 80, 168
  as present history 144–5
Gibbon, Edward 157, 165
Gladstone, William Ewart 99, 157
'Glimpse of the Domesticity of Franklyn Barnabas, A' 33
Glimpse of Reality, The 25, 54
Goethe, Johann Wolfgang von 36, 61, 122–3
Great Catherine 25, 54, 142–3
Great Man 3–4, 78, 169–70, 177
  in Shaw's plays 57–76
  see also Carlyle, Thomas; Macaulay, Thomas Babington
Grey, Lady Jane 111
Grundy, Sydney 173
Guizot, François 19
Gwynn, Nell 47

Halifax, Lord (1633–95) 3
Hallam, Henry 161–2
Harcourt, William 102
Hastings, Battle of 32, 158
Hastings, Warren 158
Heartbreak House 136, 146, 176
  as comedy and tragedy 175
  as present history 144, 158–64
Hegel, Georg Wilhelm Friedrich 161 n., 174, 176
  Shaw's knowledge of 15, 20
  and Shaw's plays 37, 38
  on World-Historical Individuals 75–6
Henri IV (France) 68
Henry II: 109
Henry IV: 80
Henry IV: (Holy Roman Empire) 109
Henry VIII: 35, 109 n., 145, 152
Herodotus 41
heroes *see* Great Man
historical facts 40–55
Hitler, Adolf 27, 36, 67, 97, 109 n., 138, 149
Hobbes, Thomas 31

Holinshed, Raphael 42
Homer 16, 159
Horthy de Nagybanya, Nikolaus 171 n.
hotels 65, 153, 155, 156
Houghton, Walter 56
Huizinga, Johan 83, 86 n.
Hume, David 7, 14, 93, 108
  Shaw's knowledge of 15
Hummert, Paul A. 19 n.
Hus, John 71
Huxley, Aldous 36–7
Hypatia 146

Ibsen, Henrik 17
Inca of Perusalem, The 66
Inge, William Ralph 171 n.
'In Good King Charles's Golden Days' 24, 25, 167, 168
  Great Man in 56, 57, 68–9
  and historical fact 45–8, 54–5
  and Macaulay's *History of England* 17, 46–7, 93–6
Irrational Knot, The 129
Irving, Henry 13, 16–17, 62
Irving, Laurence 41

James II: 95, 109, 166, 167
  see also 'In Good King Charles's Golden Days'
Jesus 36, 59, 104–5, 128
Joad, C. E. M. 36–7
John Bull's Other Island 54, 91, 162
  as comedy and tragedy 175
  hotels in 153, 155, 156
  and *Major Barbara* 156–7
  as present history 140, 144, 150–6
Jones, Henry Arthur 173
Julius Caesar 26, 42, 49, 63, 75–6, 92
  see also Caesar and Cleopatra
Justinian 27

Kaye, Julian B. vii, 162 n.
Kemal Atatürk *see* Mustafa Kemal Atatürk
Kenyon, John 24 n., 144 n.
Kéroualle, Louise de 94
Kingsley, Charles 146
Knox, John 56, 70
Kozicki, Henry 14

Lang, Andrew 37
Larson, Gale K. 42
Lassalle, Ferdinand 36

Index

Laud, William (Archbishop) 32, 112
Laurence, Dan H. 25 n.
Lawrence, T. E. 27 n.
Leibniz, Gottfried Wilhelm 36
Lenin, Vladimir Ilyich 19, 36, 67, 109, 171 n.
Leonard, Joseph 45 n.
Lincoln, Abraham 160
Louis XIV: 27, 31, 94, 96–7
Louis XVI: 66
Louis XVIII: 34
Louis Napoleon 26, 27
Lukács, Georg 169–70
Luther, Martin 7, 26, 36, 152
  see also Carlyle, Thomas

Macaulay, Thomas Babington 1–12, 14, 27, 29–30, 85 n., 92, 113 n., 115, 128 n., 158, 161
  on Carlyle 1
  on George Fox 95–6
  on Great Man 3–4, 58
  History of England 2, 9, 11, 165–7, 174–5
  and '*In Good King Charles's Golden Days*' 46–8, 93–6
  and *John Bull's Other Island* 153–6, 162
  on Middle Ages 10, 83
  on nineteenth century 9, 98–9
  on progress 4–6, 22, 100, 109, 110, 111–12, 120, 123, 129, 130, 134
  on Protestantism 2, 11
  on Revolution of 1688: 11, 98, 155
  Shaw and 15, 17–19, 32, 58 n., 81, 82, 96–9, 178
  on William III: 2, 3, 11, 166
MacDonald, Ramsay 150, 171 n.
Mahdi see Mohammed Ahmed
Mahomet 144
  and *Saint Joan* 70–1
  Shaw's unwritten play on 25–6, 47 n., 57
  see also Carlyle, Thomas
Major Barbara 36, 54, 103, 119, 146 n., 170
  as comedy and tragedy 175
  Great Man in 65–6
  as present history 140, 144, 156–8
Man and Superman 18, 70, 101, 131, 138, 146
  as comedy and tragedy 175
  progress in 116–17, 120, 121, 123–4, 125, 132, 133–4
Manetho 41
Man of Destiny, The 25
  Great Man in 57, 60, 62–3, 65, 68, 69, 70
  and historical fact 50, 53–4
  hotels in 156
Mansfield, Richard 50
Marat, Jean Paul 28–9
Marlborough, Duke of (1650–1722), 14–15
Marx, Karl 32, 36, 37, 51–2, 99–100, 148, 174
  Shaw's knowledge of 15, 18–19, 21–2
Marxism 35, 67, 78, 171
Mary Queen of Scots 14, 159
Mazzini, Giuseppe 158
Meisel, Martin vii, 17 n., 39, 55, 102, 173 n.
Merson, Luc Olivier 59
Meyerbeer, Giacomo 17
Michelangelo 45
Middle Ages 78–86, 90–1, 120, 151–2, 171, 177, 178
  see also Carlyle, Thomas; Macaulay, Thomas Babington
Mill, John Stuart 1 n., 99, 127, 140–2, 178
Millionairess, The
  hotels in 156
Milton, John 91, 124
Mirabeau, Honoré Gabriel Riqueti, comte de 57, 74
Misalliance 168
  as present history 144, 145–6, 147
Mohammed see Mahomet
Mohammed Ahmed (the Mahdi) 112
Mommsen, Theodor 24, 41, 42–3, 50, 62, 102, 104, 105
Monmouth Rebellion 46
Montaigne, Michel de 36
More, Thomas 36
Morley, John 92
Morris, William 13, 22, 44, 45, 84, 86, 98, 118, 178
  Medievalism of 78–9, 80, 81, 82, 85, 125
Motley, John Lothrop 14
Mozart, Wolfgang Amadeus 36
Mrs Warren's Profession 119
  as present history 143

Index

Murray, Gilbert 24, 43, 50, 99
Murray, T. Douglas 24–5, 44, 45 n.
Mussolini, Benito 36, 67, 97, 110, 138, 149, 171 n.
Mustafa Kemal Atatürk 36, 67

Napoleon 27, 36, 56, 57, 67, 70, 157, 159–60
  in *Back to Methuselah* 64–5
  in *The Man of Destiny* 50, 62–3, 69
  Shaw on 26, 34, 62–6
Naseby, Battle of 32
Nationalism 75, 85, 88–9, 126
Nero 43, 138
Nietzsche, Friedrich 12–13, 18
  Shaw's knowledge of 23–4
nineteenth century 9–10, 98–100
  see also Buckle, Henry Thomas; Carlyle, Thomas; Macaulay, Thomas Babington

Oates, Titus 17, 47, 95
O'Brien, James Bronterre 157
Offenbach, Jacques 102
Ohmann, Richard M. 109 n.
On the Rocks 67–8, 131
  as present history 147, 148–9
Orwell, George 27

party system 31, 93, 96–7
Passion Play (Shaw) 103
Peel, Robert 98
Penn, William 95
Perfect Wagnerite, The 79–80, 82–3, 86, 139–40
Peter the Great 41, 67
Petrie, William Matthew Flinders 114, 115–16
Philip II (Spain) 27
Pilsudski, Józef 97
Pinero, Arthur Wing 173
Pippin, King 25
Pliny 41
Plutarch 41, 42–3, 44, 61, 92
Pomponius Mela 41
Pontius Pilate 128
Powell, F. York 24
Press Cuttings 73 n., 147
Primo de Rivera, Miguel 171 n.
progress 4–9
  Shaw on 108–34, 177, 178, 179
  see also Buckle, Henry Thomas; Carlyle, Thomas; Macaulay, Thomas Babington
Protestantism 8, 89 n., 90, 119–20
  in *'In Good King Charles's Golden Days'* 94–5
  in *John Bull's Other Island* 151–3, 155
  in *Saint Joan* 75, 85–7, 89, 90, 91
  see also Carlyle, Thomas; Macaulay, Thomas Babington; Reformation

Ranke, Leopold von 40, 48–9
Reform Bill (1832) 11, 98
  see also Second Reform Bill (1867)
Reformation 11–12, 37, 74, 152, 171
  see also Protestantism
Revolution of 1688: 46, 141
  see also Macaulay, Thomas Babington
Ricketts, Charles 86 n.
Ristori, Adelaide 16
Robertson, William 14
Robespierre, Maximilien de 36, 166
Roebuck, John Arthur 147–8
Rossetti, Dante Gabriel 86 n.
Rousseau, Jean Jacques 36, 56, 149
Ruskin, John 45, 84, 98–9, 101, 125, 135, 154, 178
  Medievalism of 79, 82
  Shaw on 22, 36–7, 51–2, 97

Saint Joan 28, 37, 77–8, 81–91, 138, 144, 168
  anachronisms in 103, 107–8
  as comedy and tragedy 176–7
  Great Man in 57, 60, 69–76, 170
  and historical fact 24–5, 44–5, 47, 48, 51, 52, 55
  and *John Bull's Other Island* 151–3
  progress in 126–7
  and Tennyson's *Becket* 39
Sardou, Victorien 62–3, 172
Schiller, Friedrich 16, 47 n., 72–3
Scott, Walter 37, 103–4, 144, 165, 169
  Shaw's knowledge of 15–16
Scribe, Eugène 39, 172–3
Second Reform Bill (1867) 12, 141–2, 148
Sedley, Catherine 166
Shakespeare, William 35, 47 n., 56, 57, 78, 91
  and *Caesar and Cleopatra* 59, 60–2, 63, 92

Shakespeare, William – *cont.*
   and the chronicle play 42, 52–3
   history plays of 84–5, 98, 145, 160 n.
   Shaw's knowledge of 15, 16
Shakes versus Shav 16
Shaw, Charlotte 43
Shelley, Percy Bysshe 36
Simpleton of the Unexpected Isles, The 136, 171–2
Six of Calais, The
   and historical fact 25, 53
Sloane, William Milligan 50 n.
Smith, Adam 20
Smith, Warren Sylvester 90 n., 96
Socrates 111
Somers, Lord (1651–1716) 3
Spartacus 111
Spencer, Herbert 99
Spinoza, Baruch 36
Stalin, Joseph 36, 67, 69, 97
Stephen, Leslie 7
Stephenson, George 35
Stepniak, Sergius 49
Stock, St George 41, 42
Strabo 6, 41
Stuart-Glennie, John Stuart 23–4
Stubbs, William 90 n.
Suetonius 43
Sullivan, Barry 16
Sunderland, Earl of (1641–1702) 96–7

Tacitus 41
Taine, Hippolyte 19
Tawney, R. H. 89 n., 153
Tennyson, Alfred 13–14, 23, 122
   'Ancient Sage, The' 13
   Becket 39
   In Memoriam 121, 133 n., 168–9
   'You Ask Me Why' 7, 121
Thatcher, David S. 23 n.
Three Plays for Puritans 41, 91, 95
Thucydides 159
Tillyard, E. M. W. 84
toleration 110–12, 127–8
Torquemada, Tomás de 111
Trafalgar, Battle of 34

Trevelyan, George Macaulay 14, 24, 45, 103
Trevelyan, George Otto 41
Trojan War 158–9, 161
Trotsky, Leon 107–8
Tyler, Thomas 116, 132
Tyler, Wat 27
Tyson, Brian 45 n.

Vercingetorix 62
Voltaire, François Marie Arouet de 36, 47 n., 108, 138

Wagner, Richard 17, 36, 51, 79–80, 139–40, 170
Walpole, Robert 97, 148
Wars of the Roses 90
Waterloo, Battle of 34, 63
Watt, James 10, 35
Webb, Philip 118
Webb, Sidney 99, 178
Weber, Max 153
Weintraub, Stanley 43 n.
Wellington, Duke of 56, 63
well-made play 172–3
Wells, H. G. 36–7, 65 n., 142, 144
Wesley, John 70
White, Hayden 173–4
Whitman, Robert F. 20, 75 n.
Whitman, Walt 160
Why She Would Not 129, 156
Widowers' Houses
   as present history 143
Wilde, Oscar 27, 40, 173
Wilhelm II (Germany) 27
William III: 31, 46, 96–7
   see also Macaulay, Thomas Babington
Williams, Raymond 178
Wills, W. G. 16
Wilson, Edmund 178
World War I *see* First World War
Wyclif 71

Yeats, William Butler 115–16, 161, 169, 170, 179
   'Two Songs from a Play' 159